Y0-BSD-771

Springer Series in **Materials Science** 8

Edited by Hans-Joachim Queisser

Springer Series in **Materials Science**

Editors: U. Gonser · A. Mooradian · K. A. Müller · M. B. Panish · H. Sakaki
Managing Editor: H. K. V. Lotsch

Volume 1 **Chemical Processing with Lasers**
By D. Bäuerle

Volume 2 **Laser-Beam Interactions with Materials**
Physical Principles and Applications
By M. von Allmen

Volume 3 **Laser Processing of Thin Films and Microstructures**
Oxidation, Deposition and Etching of Insulators
By I. W. Boyd

Volume 4 **Microclusters**
Editors: S. Sugano, Y. Nishina, and S. Ohnishi

Volume 5 **Graphite Fibers and Filaments**
By M. S. Dresselhaus, G. Dresselhaus, K. Sugihara, I. L. Spain, and H. A. Goldberg

Volume 6 **Elemental and Molecular Clusters**
Editors: G. Benedek, T. P. Martin, and G. Pacchioni

Volume 7 **Molecular Beam Epitaxy** Fundamentals and Current Status
By M. A. Herman and H. Sitter

Volume 8 **Physical Chemistry of, in and on Silicon**
By G. F. Cerofolini and L. Meda

Volume 9 **Tritium and Helium-3 in Metals**
By R. Lässer

Volume 10 **Computer Simulation of Ion – Solid Interactions**
By W. Eckstein

Volume 11 **Mechanisms of High Temperature Superconductivity**
Editors: H. Kamimura and A. Oshiyama

Volume 12 **Laser Technology in Microelectronics**
Editors: S. Metev and V. P. Veiko

Volume 13 **The Semiconductor Silicon**
Materials Science and Technology
Editors: G. C. Harbeke and M. J. Schulz

G. F. Cerofolini L. Meda

Physical Chemistry of, in and on Silicon

With 26 Figures

Springer-Verlag Berlin Heidelberg New York
London Paris Tokyo

0 341-7621
CHEMISTRY

Dr. Gianfranco Cerofolini

EniChem. Via Medici del Vascello 26, I-20138 Milano MI, Italy

Dr. Laura Meda

SGS-Thomson Microelectronics, I-20041 Agrate MI, Italy

Guest Editor: Professor Dr. Hans-Joachim Queisser

Max-Planck-Institut für Festkörperforschung, Heisenbergstrasse 1
D-7000 Stuttgart 80, Fed. Rep. of Germany

Series Editors:

Prof. Dr. h.c. mult. *K. A. Müller*

IBM, Zürich Research Lab.
CH-8803 Rüschlikon, Switzerland

Prof. Dr. *U. Gonser*

Fachbereich 12/1
Werkstoffwissenschaften
Universität des Saarlandes
D-6600 Saarbrücken, FRG

A. Mooradian, Ph. D.

Leader of the Quantum Electronics Group, MIT,
Lincoln Laboratory, P.O. Box 73,
Lexington, MA 02173, USA

M. B. Panish, Ph. D.

AT&T Bell Laboratories,
600 Mountain Avenue,
Murray Hill, NJ 07974, USA

Prof. *H. Sakaki*

Institute of Industrial Science,
University of Tokyo,
7-22-1 Roppongi Minato-ku,
Tokyo 106, Japan

Managing Editor:

Dr. Helmut K. V. Lotsch

Springer-Verlag, Tiergartenstrasse 17
D-6900 Heidelberg, Fed. Rep. of Germany

ISBN 3-540-19049-X Springer-Verlag Berlin Heidelberg New York
ISBN 0-387-19049-X Springer-Verlag New York Berlin Heidelberg

Library of Congress Cataloging-in-Publication Data. Cerofolini, G. (Gianfranco), 1946- Physical chemistry of, in and on silicon / C. [i.e. G] Cerofolini, L. Meda. p. cm–(Springer series in materials science ; v. 8) Bibliography: p. Includes index. ISBN 0-387-19049-X (U.S.) 1. Silicon. I. Meda, L. (Laura), 1957-. II. Title. III. Series. QD181.S6C48 1989 546'.683–dc 19 89-4244

This work is subject to copyright. All rights are reserved, whether the whole or part of the material is concerned, specifically the rights of translation, reprinting, reuse of illustrations, recitation, broadcasting, reproduction on microfilms or in other ways, and storage in data banks. Duplication of this publication or parts thereof is only permitted under the provisions of the German Copyright Law of September 9, 1965, in its version of June 24, 1985, and a copyright fee must always be paid. Violations fall under the prosecution act of the German Copyright Law.

© Springer-Verlag Berlin Heidelberg 1989
Printed in the United States of America

The use of registered names, trademarks, etc. in this publication does not imply, even in the absence of a specific statement, that such names are exempt from the relevant protective laws and regulations and therefore free for general use.

2154/3150-543210 – Printed on acid-free paper

Preface

QD 181
S6
C481
1989
Chem

The aim of this book is twofold: it is intended for use as a textbook for a course on electronic materials (indeed, it stems from a series of lectures on this topic delivered at Milan Polytechnic and at the universities of Modena and Parma), and as an up-to-date review for scientists working in the field of silicon processing. Although a number of works on silicon are already available, the vast amount of existing and new data on silicon properties are nowhere adequately summarized in a single comprehensive report. The present volume is intended to fill this gap.

Most of the examples dealt with are taken from the authors' everyday experience, this choice being dictated merely by their greater knowledge of these areas. Certain aspects of the physics of silicon have not been included; this is either because they have been treated in standard textbooks (e.g. the inhomogeneously doped semiconductor and the chemistry of isotropic or preferential aqueous etching of silicon), or because they are still in a rapidly evolving phase (e.g. silicon band–gap engineering, generation–recombination phenomena, cryogenic properties and the chemistry of plasma etching).

In line with the standard practice in microelectronics, CGS units will be used for mechanical and thermal quantities, and SI units for electrical quantities. All atomic energies will be given in electronvolts and the ångstrom will be the unit of length used for atomic phenomena.

The symbols '$\approx$' and '$\simeq$' are used with the following meanings:
$a \approx b$ means 'a is of the same order as b', say $\frac{1}{3}b < a < 3b$; this relation usually follows from experimental findings or from estimates of orders of magnitude;
$a \simeq b$ means 'a is approximately equal to b'; this relationship is stronger than $\approx$ ($= \Rightarrow \simeq \Rightarrow \approx$) and often follows from the replacement of an asymptotic relation ($\sim$) by the finite relation. If for instance $x \to x_0 \Rightarrow f(x) \sim g(x)$ we shall often assume that $|x - x_0| < \delta \Rightarrow f(x) \simeq g(x)$, where δ is not required to vanish.

As far as we are aware, the literature list, although not exhaustive, is up to date. When a reference is quoted in the text we are not necessarily attributing priority of discovery to the authors cited – we have simply chosen a sufficiently modern and comprehensive account of the relevant subject.

Any suggestions from readers as to how we moght update the book will be much appreciated and may well be useful for a subsequent edition (which will surely be necessary if silicon remains the most important material for electronic devices).

Finally, we wish to thank our friends and colleagues who read and commented on various drafts of this work: Drs A. Armigliato, M. Servidori and S. Solmi (Istituto LAMEL, Bologna). Moreover, special thanks are due to L.I. Emanuelli for his precious help. We are also deeply indebted to Dr. A.M. Lahee (Springer-Verlag) for her editorial assistance and for her help in rendering our poor English more English.

Milano, August 1988

G.F. Cerofolini
L. Meda

Contents

1. Silicon

The group IV elements furnish us with perhaps the best example of the gradual gain in metallic character with the increase of the atomic number Z. Evidence for this gain can be found in both the atomic and condensed-phase properties.

For instance, the first ionization energy E_{ion}, the electronegativity ξ and the cohesive energy U_0 decrease rather regularly from carbon to lead, while the tetrahedral radius r increases (Table 1.1).

Table 1.1: First ionization energy, cohesive energy, electronegativity and tetrahedral radius of group IV elements

Element	E_{ion}[eV]	U_0[eV]	ξ	r[Å]
C	11.260	7.37	2.5 – 2.6	0.77
Si	8.151	4.63	1.8 – 1.9	1.17
Ge	7.899	3.85	1.8 – 1.9	1.22
Sn	7.344	3.14	1.8 – 1.9	1.40
Pb	7.416	2.03	1.8	1.44

The data for electronegativity are somewhat controversial, and those reported here are taken from the discussion of *Cotton* and *Wilkinson* [1.1]. The gradual gain in metallic character may also be understood by considering the crystalline structures formed by these elements. Indeed, from carbon to germanium the stable and metastable phases are the ones typical of purely covalent solids; lead has the typical crystalline structure of metals; while tin has two phases, one typical of covalent materials and the other with intermediate character (Table 1.2).

For the elements admitting the diamond-cubic structure, the electronic properties evolve quite regularly with Z from the ones typical of an insulator (carbon) to the ones characteristic of a metal (tin). This can be seen by comparing, for instance, the energy gap E_g and the dielectric constant ε (Table 1.3).

Table 1.2: Crystalline structures of stable and metastable phases of group IV elements

Element	Phases			
C	graphite	diamond cubic		
Si		diamond cubic		
Ge		diamond cubic		
Sn		diamond cubic	β-tin	
Pb				face-centred cubic

Table 1.3: Energy gap and dielectric constant of group IV diamond-cubic crystals (0 K)

Element	E_g[eV]	ε
C	5.4	5.5
Si	1.17	11.8
Ge	0.74	15.8
α-Sn	0.0	

1.1 Elemental Silicon

Silicon is one of the most abundant elements in the Earth's crust, occurring mainly in the form of SiO_2 (quartz, etc.) and silicates. The analysis of lunar samples (for a total amount of 381.69 kg, supposedly representative of the whole of the Moon's crust) taken in Apollo missions, showed that silicon ist also one of the most abundant elements in the Moon's crust. The Rutherford backscattering spectroscopy (RBS) analyses carried out during the Viking missions confirmed the abundant presence of silicon in Mars too [1.2]. The presumed relative atomic abundance of silicon in the Universe is 6×10^{-5}.

The atomic weight of silicon is 28.09 g/mol; this is because natural silicon is a mixture of the isotopes ^{28}Si (relative abundance = 92.3 %), ^{29}Si (4.7 %) and ^{30}Si (3.0 %). Natural silicon has a very weak radioactivity ($\approx$ 1 decay/hr$\times$g) due to the isotope ^{32}Si.

From the electronic point of view, silicon has an sp^3 hybridization and forms tetrahedrally-coordinated compounds with both oxygen and hydrogen (Fig. 1.1):

Silicon is usually found in the oxidized form SiO_2 because of the high negative free energy of the dioxide; SiO_2 is one of the most stable oxides, as shown by Table 1.4 which is filled in order of increasing stability of the metal-oxygen bond. This table gives the formation free energy ΔF_0 of the most stable oxides and this energy per oxygen atom [1.3].

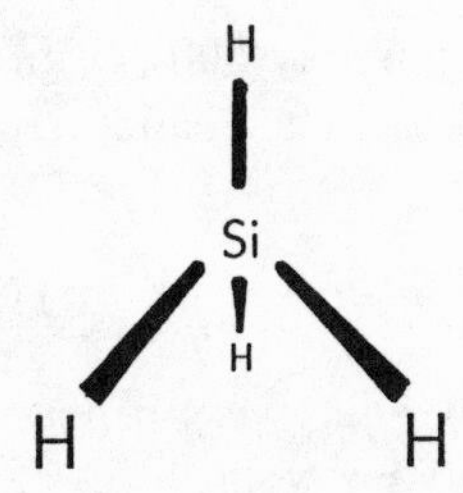

Fig. 1.1 Pictorial view of the SiH_4 molecule

Table 1.4: Free formation energy of the most stable oxides

Oxide	$-\Delta F_0$[kcal mol^{-1}]	$-\Delta F_0$[kcal g at^{-1}]
NiO	46.1	46.1
MoO_2	114.5	57.2
Cr_2O_3	240.2	80.1
Ta_2O_5	434.9	87.0
SiO_2	187.9	94.0
TiO	112.2	112.2
Al_2O_3	362.1	120.7

The Si-O bond is moderately polar, having a Pauling electronegativity difference of 1.54; this corresponds to a 33% ionic character.

Because of its similarity to carbon, a kind of silicon 'organic chemistry' can be hypothesized; however, the Si-Si and Si-H bonds are much weaker than the C-C and C-H bonds, respectively, and silicon has very little tendency to form π bonds, so that only few silicon 'organic compounds' are known; among them silane SiH_4 and disilane SiH_3-SiH_3.

Of particular interest are the chlorosilane compounds, from SiH_3Cl to $SiCl_4$, which are intermediates in the purification of metallurgical silicon.

1.2 Silicon Metallurgy

Silicon is obtained in elemental form by the reduction of purified SiO_2 with carbon or CaC_2 at about 1700 °C:

$$SiO_2 + C \rightarrow Si + CO_2 \uparrow \ .$$

After reduction, the silicon is in the form of a fine-grain powder and contains a lot of impurities.

Up to now the preparation of ultrapure silicon appears impossible without the transformation of this element into gaseous compounds. The industrial process consists of the transformation in a fluid bed reactor of silicon into chlorosilane compounds at a temperature of approximately 300 °C. In these conditions the compound produced with maximum yield is trichlorosilane:

$$Si + 3\,HCl \rightleftharpoons SiHCl_3 + H_2 \ . \quad (1.1)$$

The low boiling point of $SiHCl_3$ (31.8 °C) allows a very effective purification by means of fractioned distillation due to comparably low volatilities of practically all possible impurities. Silicon is then obtained by high temperature ($\simeq$ 900 °C) reduction of $SiHCl_3$ in a hydrogen atmosphere by means of the reverse of reaction (1.1). The deposited silicon results in high purity, high resistivity, polycrystalline rods which are the starting point for the preparation of the single crystal.

1.3 Single-Crystal Growth

Silicon single crystals can be obtained as ingots with a diameter of 20 cm and a length of 300 cm; they are surely among the largest single crystals ever produced. Their industrial production is based on two techniques — the Czochralski (CZ) method and the float-zone (FZ) method. These methods are briefly described by *Herrmann* et al. [1.4]; more details on the CZ method are given in a review of *Zulehner* and *Huber* [1.5].

Czochralski Technique The CZ method was originally developed for metals and, since its application to silicon, larger and larger dislocation-free silicon single crystals have been grown.

In this method, the ultrapure polycrystalline silicon is melted with the addition of the required amount of dopant in a quartz crucible maintained in vacuum or in an inert atmosphere. A seed, i.e. a piece of silicon single crystal of the required orientation, a few millimetres in diameter, is added to the molten silicon and slowly withdrawn from it, while both the crucible and the growing crystal are rotated. The pulling rate, rotation speed, power input, etc. are adjusted to achieve the desired crystal diameter.

The impurities taken up from the crucible walls do not allow CZ silicon to be obtained with electrical resistivity higher than 100 Ω cm (p-type material) or 40 Ω cm (n-type). In addition, the high oxygen content of CZ silicon (see Chap. 4) can cause severe resistivity changes during wafer processing, because of the activation of the oxygen as a donor. These drawbacks make the CZ material unsuitable for large area, high voltage devices. For these applications the FZ material is necessary.

Float-Zone Technique In the FZ growth method a polycrystalline rod is locally melted by an induction coil; the molten zone, which remains floating because of surface tension, is forced to travel from one end of the rod to the other and impurities segregate preferentially in one of the two phases (solid and liquid) in contact. Most impurities (in practice all except oxygen) preferentially segregate into the liquid. If the first end is melted together with a seed, the solidified silicon will be a single crystal with the same crystalline orientation as the seed. In both CZ and FZ growth methods, as the crystal is pulled or the molten zone is moved at a speed f, it is also rotated around an axis which is generally different from the one of thermal symmetry.

Temperature fluctuations, experienced by the solid-liquid interface, result in fast modulations of the growth rate, from values of f much higher than the average value to zero or even to negative values (remelting).

As a consequence, the effective segregation coefficient K can be very different from the equilibrium value K_0, due to the limited supply of impurity through the boundary layer, of thickness d at the growing interface.

The following expression

$$K = \frac{K_0}{K_0 + (1 - K_0)\exp(-fd/D)} ,$$

where D is the impurity diffusion coefficient in the melt, shows the correct limiting values for negligible and very-high growth rates, respectively: $f \to 0 \Rightarrow K \to K_0$, and $f \to \infty \Rightarrow K \to 1$ [1.6]. Impurities with very low coefficient K_0 will present strong concentration fluctuations, which in turn cause the formation of the resistivity striations, i.e. of marked variations of resistivity along the ingot diameter.

Values of K_0 for a number of impurities are reported in Table 1.5.

Table 1.5: Equilibrium segregation coefficient of impurities between solid and liquid silicon

Element	K_0	Element	K_0
Li	1.0×10^{-2}	As	3.0×10^{-1}
Cu	4.0×10^{-4}	Sb	2.3×10^{-2}
Ag	1.0×10^{-6}	O	1.4
Zn	1.0×10^{-5}	Cr	2.8×10^{-3}
B	8.0×10^{-1}	Ti	9.0×10^{-6}
Al	2.0×10^{-3}	Fe	8.0×10^{-6}
C	6.0×10^{-2}	Co	8.0×10^{-6}
P	3.5×10^{-1}	Ni	3.0×10^{-5}

The FZ material can be produced with lower impurity content than the CZ material, but tends to form concentration striations, and these fluctuations are particularly harmful to certain families of electron devices (thyristors, large area transistors, rectifiers).

The high resistivity FZ silicon is usually p-type doped, because of the equilibrium segregation coefficient of boron. An increase of resistivity can be obtained by compensation, and to achieve that neutron transmutation doping of silicon is used. The neutron transmutation doped material is obtained from the FZ material by the $^{30}\mathrm{Si}(\mathrm{n},\beta)\,^{31}\mathrm{P}$ reaction followed by a heat treatment to anneal the radiation damage. The above reaction is obtained by placing the ingot in suitably developed nuclear reactors. The low absorption cross-section of thermal neutrons makes it possible to obtain very uniformly doped crystals with diameter larger than 100 mm.

1.4 Mechanical Properties

For the theory of the elastic properties of cubic crystals we refer the reader to [1.7]. Here we are mainly interested in two quantities which will be useful in the following — the fracture limit and the plastic limit.

1.4.1 Fracture Limit

This quantity is usually defined in a macroscopic framework; from the microscopic point of view, the fracture limit τ_f is an ill-defined quantity. Indeed, silicon can absorb a huge stress provided that the total energy involved in it is small enough. So a single atom can deform the crystal in its neighbourhood by an amount as large as 20% without the formation of microcracks. If however the total energy involved in the stress is high, a fracture can occur — the more extended the stressed region, the lower the fracture limit.

In spite of this ill-definition, we assume that a fracture limit can be defined and we assume for it, at least tentatively, the value given by *Sylvestrowicz* [1.8]. The graph of τ_f in the temperature range $0-600$ °C is shown in Fig. 1.2.

1.4.2 Plastic Limit

When the temperature is high enough, a high external stress can be absorbed by the crystal without fracture through the formation of dislocations. Each dislocation is able to store an energy excess in the crystal of the order 10 eV per crossed plane, which excludes that dislocations are equilibrium defects; dislocations are typically formed when a macroscopic energy is released in a small portion of the crystal.

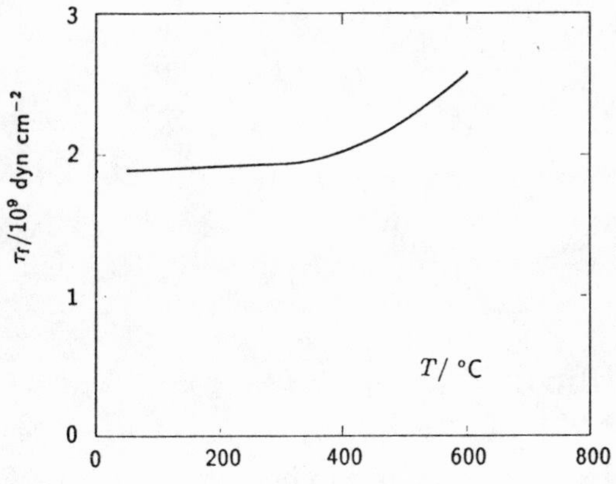

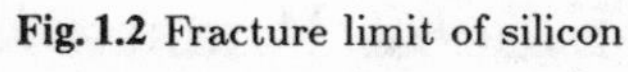
Fig. 1.2 Fracture limit of silicon

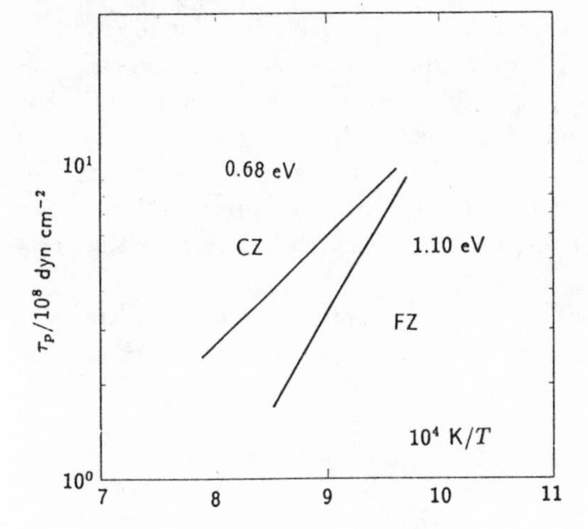

Fig. 1.3 Plastic limit of CZ and FZ silicon; the activation energies to form dislocations are reported

The plastic limit τ_p of the FZ material is different from that of CZ material, as Fig. 1.3 [1.9] shows. This comparison shows that the plastic limit of the CZ crystal is higher than that of the FZ crystal, the relative difference being greater the higher the temperature. Since dislocations cause severe deviations from ideal behaviour of electronic devices, the previous comparison suggests the use of CZ crystals at least in semiconductor device manufacturing. The difference between FZ and CZ materials is ascribed to the pinning action on dislocation by oxygen rather than to a change of τ_p itself [1.10]. Indeed, in dislocation-free CZ material the *initial* plastic yield is close to that of FZ material.

2. Silicon Phases

2.1 Diamond-Cubic Silicon

Up until to a few years ago, silicon was presumed to have only one crystal structure — the cubic diamond structure [2.1]. The diamond-like structure of silicon is face-centred cubic (f.c.c.) with two atoms per unit cell; the lattice constant a is 5.431 Å and the atomic density N_{Si} is 5.0×10^{22} cm^{-3}.

From the topological point-of-view, in diamond-cubic (d.c.) crystals the ring (i.e. the closed path connecting neighbouring atoms) of minimum size is six membered, not planar and of the 'chair' type. Because of the impossibility of representing in a plane six-fold rings where each vertex has a tetrahedral coordination, our two-dimensional (2D) representation of the d.c. lattice will be a square lattice, in such a way as to preserve the stoichiometry.

The energy gap E_{g} is 1.17 eV at 0 K and charge screening effects take place via a large relative dielectric constant $\varepsilon_{\mathrm{Si}}$ ($\varepsilon_{\mathrm{Si}} = 11.8$ in the static limit).
Extensive data for mechanical, thermal, optical and electrical properties of silicon are reported in *Wolf's Silicon Semiconductor Data* [2.2].

2.2 Diamond-Hexagonal Silicon

Recent experiments on hydrostatically compressed d.c. silicon have however shown the existence of other two phases: the β-tin and simple hexagonal. The existence regions and the energies of these phases can be determined theoretically from quantum mechanical calculations [2.3–5]. Excellent agreement exists between experimental findings and theoretical predictions.

For the wurtzite-like hexagonal phase of silicon, henceforth referred to as diamond hexagonal (d.h.) silicon, which is theoretically described as a metastable phase with a small energy excess (≈ 0.01 eV/atom), experimental evidence has been found for anvil indentation at temperatures in the range 400 – 700 °C (with maximum intensity around 550 °C [2.6]), and for ion implantation in the high current, high temperature (> 200 °C) mode [2.7–10]. The agglomerates of self-interstitials lying on {113} planes, which typically form after ion implantation at room temperature [2.11], have been postulated to be the diamond hexagonal phase of silicon [2.12].

The results concerning anvil indentation can be explained by assuming that the diamond hexagonal phase is stable only for stresses higher than a transition value τ_{dc-dh} approximately equal to 2×10^9 dyn/cm^2. For temperatures below 400 °C this stress is higher than the fracture limit of silicon and therefore produces microcracks, while for temperatures higher than 700 °C it is plastically absorbed by the formation of dislocations. Figure 2.1 shows the suggested existence region of the d.h. phase in the stress-temperature plane.

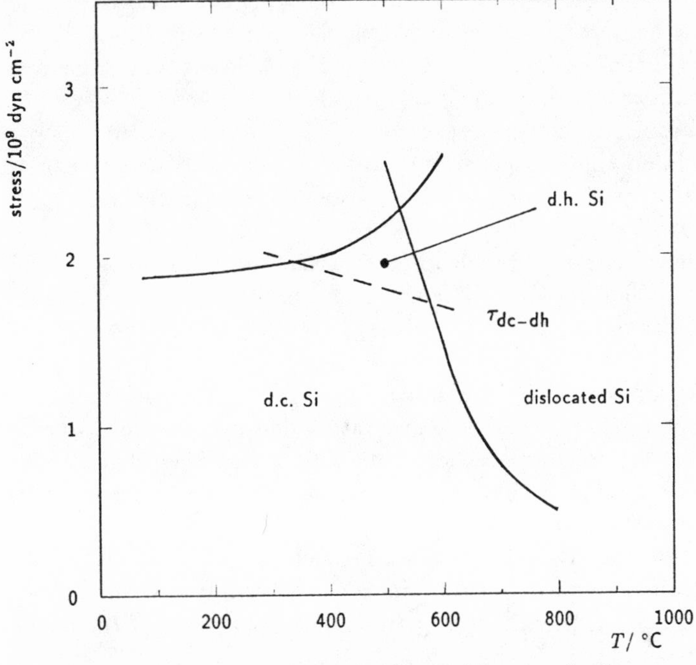

Fig. 2.1 Suggested existence region for diamond hexagonal silicon

The fact that the diamond hexagonal phase is obtained in ion-implanted silicon at macroscopic temperatures lower than the one at which this phase starts to form during anvil indentation, suggests that during the implantation there is a local heating sufficient to bring the silicon into the existence region of the metastable phase. An existence domain in Kr^+ implanted samples in terms of less fundamental (but more operational) parameters, i.e. implanted fluence and target temperature, is reported in [2.10].

2.3 Amorphous Silicon

Amorphous silicon can be obtained by different preparation techniques — glow discharge, evaporation, chemical vapour deposition, or ion implantation. The properties of the layer depend strongly upon the preparation technique. In this work we shall mainly consider the amorphous silicon (a Si) which can be obtained by ion implantation (I^2) of d.c. silicon.

Prior to considering the possible kinds of amorphous layers which can be obtained by ion implantation, we give a short discussion of the mechanisms through which the ion (usually with an atomic weight A in the range 10–100 and energy E in the range 20–200 keV) loses its energy. Energy loss takes place by collisions with electrons (electronic energy loss) and with atoms (nuclear energy loss). Energy lost by collisions with electrons is quickly transformed into heat, while experiments carried out by implanting relatively light ions (boron) at energy in the range $20 - 40$ keV suggest that the energy lost in collisions with atoms is responsible for three phenomena:
1) when the collision has a high impact parameter, the transferred energy is low (< 40 eV) and the displaced target atom transfers its energy excess to the neighbouring ones in the form of vibrations, so that lattice regularity is eventually preserved;
2) when the collision has a fairly low impact parameter, the transferred energy is moderately high ($0.04 - 1$ keV) and the recoiled atom may remain permanently off-site, so originating a vacancy-interstitial pair (Frenkel defect);
3) when the collision has a very low impact parameter, the transferred energy is high enough (say, > 1 keV) to generate a dense collisional cascade; the collisional cascade originates a hot cloud which, if sufficiently hot (i.e., for transferred energy higher than 5 keV), results after quenching in an island of displaced atoms.

We believe that the displaced atoms in such islands (which eventually overlap to form a continuous layer) form the ion-implanted amorphous phase. Experimental evidence and a quantitative description of the above view are presented in [2.13,14].

2.3.1 Amorphous 1 (a_1)

In most situations (i.e. for implantation energy much higher than the one necessary to form the hot cloud, and for ions not lighter than boron) the number of atoms which remain displaced because of mechanism 3) can be evaluated simply by considering the deposited energy. This quantity varies with depth and with the implant energy, and its maximum is at a fraction of about 80% of the projected range R_p of the ion in silicon [2.15,16].

For displacement events independent of one another, the expected number of displaced atoms per unit area N_{dis} should increase as

$$N_{dis} = N_{\infty}[1 - \exp(-\Phi/\Phi_c)] \quad ,$$

where Φ is the implanted fluence, N_{∞} is the number of displaced atoms immediately after the formation of a continuous amorphous layer, and Φ_c is a characteristic fluence; roughly speaking Φ_c is the amorphization fluence, i.e. the fluence at which a continuous amorphous layer is formed. Some investigations of the damage release show a superlinear increase of N_{dis} with Φ, this phenomenon being more evident the lighter the implanted ion [2.17].

This superlinear increase is easily understood by observing that it is easier to amorphize an already damaged crystal than an undamaged one [2.14].

The characteristic fluence Φ_c weakly depends on the implantation energy E (provided that E is high enough), but very strongly upon ion mass and target temperature. Table 2.1 collects some experimental values of Φ_c for boron, phosphorus, arsenic and antimony projectiles in silicon at room temperature.

Table 2.1: Amorphization fluence for B, P, As and Sb

Ion	E[eV]	R_p[nm]	Φ_c[cm^{-2}]	Reference
B	25	82	9×10^{15}	[2.14]
P	90	110	3×10^{14}	[2.18]
As	150	85	8×10^{13}	[2.14]
As	190	106	8×10^{13}	[2.18]
Sb	80	38	1×10^{14}	[2.14]

Once a thin layer is completely amorphized, any increase of implanted fluence is responsible for an extension of the damaged region both in depth and toward the surface.

Consider now an implantation at a fluence just above Φ_c; in this situation a layer of thickness x_{dis} centred before R_p is amorphous, and the amount of vacancy - self-interstitial (v–i) pairs created is given by $y_{vi}\Phi_c$, where y_{vi} is the yield for such events. Taking this yield to be of the order of 10, and considering that such defects are spread over a region of the order of $2R_p$ [2.19], provided that the implanted ion is heavy enough (e.g., As), Φ_c is low, the concentration of vacancies and self-interstitials is small compared with silicon atomic density N_{Si} (say, 10^{19} cm^{-3} compared with 5×10^{22} cm^{-3}), and the energy stored in the crystal is insufficient to allow it to revert to the amorphous phase. In this situation the amorphous appearence cannot be ascribed to point-like defects continuously built-up in the crystal during the implantation and must therefore be thought of as due a single event displacing a lot of atoms from their lattice location. We think that in this amorphous phase, referred to as a_1, the d.c. topological order is preserved, i.e., bonds are distorted but most of them not broken. Evidence for this is presented in [2.20].

That the amorphous phase obtained by ion implantation retains memory of the crystalline phase is also suggested: by electroreflectance studies, showing that one specific structure (E_1) of the interband spectrum of silicon (evidence for crystal order) survives up to complete amorphization [2.21], and by multiple-crystal X-ray diffraction studies, showing that the ion-implanted amorphous phase can be thought of as containing vacancies and self-interstitials thus providing evidence for a topological order [2.22].

Preservation of bonds, though distorted, is favoured by the strongly covalent nature of silicon, and the relative weight of amorphization compared to v–i pair creation depends upon mass and energy of the impinging atom. If the mass of the impinging atom is high enough (e.g., arsenic in silicon), crystalline order can be destroyed preserving, however, the topological order.

The transition from the a_1 phase to d.c. does not require bond breakdown but simple atom rearrangement, so that the a_1 layer can be reconstructed by heat treatment at moderate temperatures (> 450 °C). The reconstruction process, known as solid phase epitaxy (SPE) and starting at the undamaged bulk seed, is thermally activated with an activation energy of about 2.5 eV and a pre-exponential factor of the order of the sound velocity [2.23]. This coincidence upholds the idea that the topological order in a_1 Si is the same as in d.c. Si.

The transformation properties of the a_1 phase can be summarized in the following scheme:

$$\text{d.c.} \overset{\text{low } T \text{ I}^2}{\longrightarrow} \text{a}_1 \overset{>450^\circ\text{C SPE}}{\longrightarrow} \text{d.c.}$$

The electronic properties of the amorphous phase obtained by ion implantation, determined by electron spin resonance and refractive index measurements, are quite independent of the implanted ion in terms of atomic weight (from carbon to tin), implantation energy (in the range 0.2 – 2 MeV), and fluence (in the range $10^{16} - 10^{17}$ cm^{-2}) [2.24]; this feature allows us to think of the a_1 layer as a ‘true’, rather than a technological, process-dependent, structure.

When the fluence increases, the number of v-i pairs produced increases monotonically with Φ, eventually reaching a value not negligible with respect to N_{Si} (some percent, say). This situation is characterized by a vacancy excess close to the surface and a self-interstitial excess at greater depth, which can reorganize to form extended defects provided the temperature is high enough. Evidence for differently reconstructed amorphous layers obtained by ion implantation, presumably to be ascribed to different self-interstitial excess, was given in [2.25].

If the implantation takes place on substrates kept at relatively high temperature, vacancies and self-interstitials generated in pairs during the implantation migrate, partially recombine or form extended defects, and displaced atoms in the amorphous phase return to their equilibrium positions [2.26]. Implantations at these temperatures therefore result in the formation of partially-reconstructed silicon, the residual damage being formed by dislocation segments, twins, stacking faults and rod-like diamond hexagonal defects [2.7-10]. This damage does not reconstruct by SPE; however, at temperatures above 1100 °C, strongly-damaged silicon melts and the crystal can be reconstructed by liquid phase epitaxy starting at the undamaged bulk.

2.3.2 Amorphous 0 (a_0)

An amorphous layer obtained by melting silicon and quenching it, without an underlying crystalline seed, is completely different from the amorphous phase obtained by ion implantation. In fact, this sudden quenching should lead to a state of supercooled liquid with octahedral coordination and metallic character. This phase is the one which is more easily modelled by current theories.

Amorphous phase with semiconductor properties can be obtained by maintaining the tetrahedral coordination [2.27,28]. An amorphous phase with this coordination and with no topological order, referred to as a_0, can be obtained by glow discharge from SiH_4 [2.29]; this phase is characterized by a dangling bond density of the order of 10^{19} cm^{-3} which can be passivated by reaction with hydrogen.

Table 2.2 summarizes a few thermodynamic properties of the amorphous phases compared with those of the stable crystalline phase.

Table 2.2: Thermodynamic properties of silicon phases

Phase	Stability[°C]	Energy excess[eV]	Order
d.c.	1417	0	crystalline
a_1	450	0.13	topological
a_0	$\simeq$ 1100	0.12	none

The values for the energy excess of the a_1 phase are taken from [2.30,31] and for the a_0 phase from [2.32].

A completely different view of amorphous silicon comes from recent high resolution electron microscopy (HREM) and localized electron diffraction studies, suggesting a submicrocrystalline model for the structure of non-crystalline solids. In these studies [2.33–35] Bragg diffraction from regions of amorphous silicon (in the length scale above 10 nm) reveals the existence of crystalline clusters too small to be revealed by X-ray diffraction in bulk, and the clusters are localized by HREM.

3. Equilibrium Defects

Equilibrium defects are lattice defects whose existence is thermodynamically guaranteed by the finite non-zero temperature of the lattice. These defects are therefore characterized by an equilibrium concentration.

Though some equilibrium defects disappear in the thermodynamic limit (e.g., surfaces), however they can play a fundamental role in determining the concentration of defects which do not disappear in the thermodynamic limit (e.g., vacancies, self-interstitials and their clusters). In fact, a mechanism through which equilibrium defects are generated is just surface reconstruction.

For a process at constant pressure, the equilibrium atomic density N_x of the point defect x (x = v, vacancy; i, self-interstitial;...) at absolute temperature T is given by

$$N_\mathrm{x}/(N_\mathrm{x}^0 - N_\mathrm{x}) = \exp(-\Delta G_\mathrm{x}/k_\mathrm{B}T) \quad , \tag{3.1}$$

where N_x^0 is the atomic density of allowed sites for the defect x, ΔG_x is the atomic formation Gibbs free energy of the defect and k_B is the Boltzmann constant. The number of allowed sites depends on the nature of the defect. For the vacancy, $N_\mathrm{v}^0 = N_\mathrm{Si}$, and presumably in all cases $N_\mathrm{x}^0 \approx N_\mathrm{Si}$.

The same equation, (3.1), holds true for a process at constant volume, provided that ΔG_x is replaced by the Helmholtz free energy ΔF_x.

These considerations hold true only for processes where all defects can be absorbed or injected independently of one another; in particular, they hold when surface reconstruction is possible. However, in most practical situations the sample surface is subjected to external constraints and, therefore, is not free to self-reconstruct. An example of such a constraint is an oxide layer, which prevents surface reconstruction because of the strength of the Si-O bonds.

Thus, as well as the processes at constant pressure or volume usually considered by thermodynamics, for equilibrium defects one must also consider the *process at constant surface.* Not only does this process keep constant the total area, but also leaves unchanged the shape, i.e. the distribution of various surfaces of area $A_{[hkl]}$ and orientation $[hkl]$, $\forall h,k,l : A_{[hkl]} =$ constant. Such a process is characterized by the condition

$$\Delta N_\mathrm{v} = \Delta N_\mathrm{i} \tag{3.2}$$

where ΔN_x is the number of defects (x = v, i) injected by the process in addition to pre-existing ones. Condition (3.2) follows from the fact that, in absence of surface reconstruction, vacancies and self-interstitials are necessarily generated in pairs; it holds true not only in equilibrium conditions, but also for all processes where surface reconstruction is not allowed.

We feel that some discrepancies in data on vacancies and self-interstitials from different authors are actually due to neglect of the influence of constraints forbidding surface reconstruction, an example being the native oxide layer usually present because of high silicon reactivity.

3.1 Vacancies

A vacancy is a lattice site associated with a missing atom. The missing atom leaves four dangling bonds which can form two molecular orbitals. In this case (neutral vacancy, v) different electronic configurations are possible:

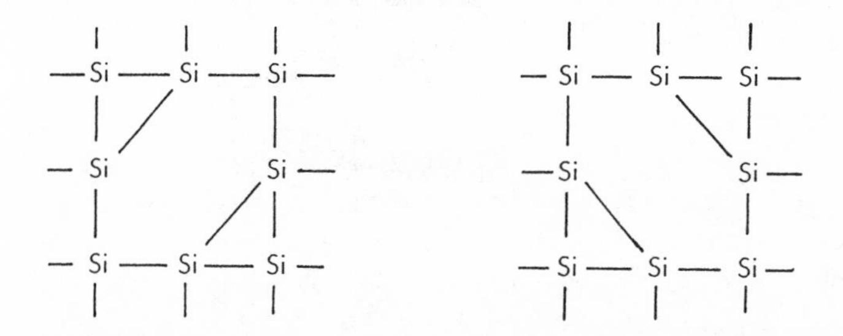

There is thus the possibility that the electronic degeneracy be lifted through relaxation to a more stable configuration by a Jahn - Teller distortion [3.1]. The major difficulties in the theoretical analysis of equilibrium defects come from relaxation phenomena, which might involve a significant number of lattice atoms. The 'extended nature' of point-like defects, however, is still an open question (see next two sections).

The molecular orbitals formed by the superposition of dangling bonds are obviously strained and can be broken with relative ease by the interaction with electrons or holes from the conduction or valence band, respectively. After the interaction, the vacancy acquires a positive or negative charge, e.g.

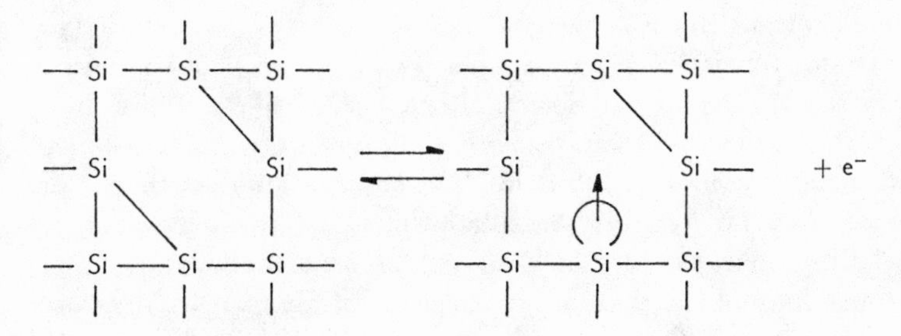

or, more schematically,

$$v \rightleftharpoons v^{+} + e^{-} \quad .$$

Four charge states of the vacancy have been hypothesized; these states and their trapping behaviour are reported in Table 3.1, though charge state assignement of the electronic levels of the vacancy is still an unsettled question [3.2]. Even the existence of the positive doubly-ionized vacancy is controversial.

Table 3.1: Charge states and trapping behaviour of the vacancy

Charge state	Trapping behaviour
v	donor (0/+)
v^{-}	acceptor (−/0)
$v^{=}$	acceptor (− − /−)
v^{+}	

The equilibrium concentration of vacancies in silicon is much lower than in a metal. The qualitative reason for this fact is the following: the packing density of the diamond lattice is only 34 % compared to 74 % for a closed-packed structure such as the f.c.c. lattice of metals (e.g., copper and silver). Because of this geometrical condition, one expects that the incorporation of interstitial atoms in silicon is favoured, while the formation of vacancies occurs with difficulty. In passing, it is worthwhile noting that the very open structure of the d.c. lattice makes the a_1 configuration possible.

For vacancies, $N_v^0 = N_{Si}$; this property, however, is the only one about which there is a general agreement among different authors. For instance, decomposing ΔG_v into its enthalpy and entropy contributions, $\Delta G_v = \Delta H_v - T\Delta S_v$, *Masters* and *Gorey* [3.3] give $\Delta H_v = 3.66$ eV, $\Delta S_v = 7k_B$; the most recent theoretical calculations do not estimate the formation entropy and give a formation enthalpy of about 5 eV [3.4]. The high experimental value of entropy suggests that the vacancy has an extended nature. The problem of whether or not this is actually the case, has long been debated in the literature. For instance, in an early paper *Van Vechten* [3.5] suggested that the vacancy is a true point-like defect with a local octahedral morphology; in a recent paper, however, *Van Vechten* [3.6] postulates the existence of extended vacancies; *Seeger* and *Frank* [3.7], however, state that the latter model of the vacancy is not tenable to explain diffusivity in silicon.

We shall briefly discuss the idea of Van Vechten on extended vacancies. Since in the amorphous phase the energy excess is of about 0.1 eV/atom, configurations with several (say, up to 40) displaced atoms should be energetically equiprobable with a single vacancy; such amorphous clusters could then arrange to form extended vacancies or self-interstitials. The latter possibility can actually occur only if topological order in the amorphous cluster is not preserved, in contradiction to the proposed view of the a_1 phase.

3.2 Self-Interstitials

No generally accepted model of the silicon self-interstitial is known. We recall here only the early view of the self-interstitial as an almost free silicon atom embedded in the silicon matrix [3.8]

```
    |         |
  — Si —————— Si —
    |         |
    |    Si   |
    |         |
  — Si —————— Si —
    |         |
```

and the more recent view of the self-interstitial as a pair of silicon atoms in a dumb-bell configuration upon a lattice site with a ⟨110⟩ or ⟨100⟩ orientation [3.9]:

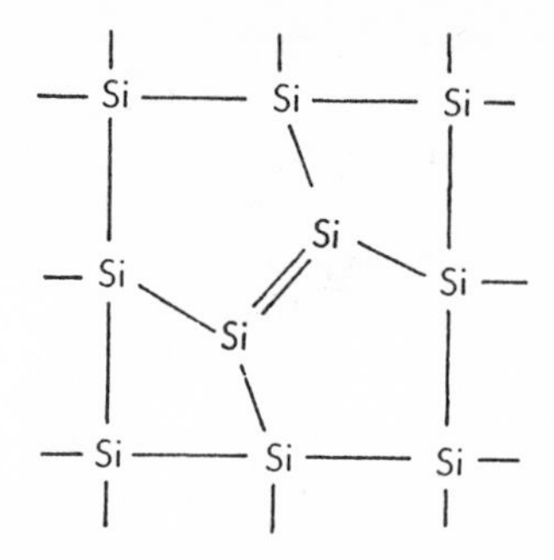

In the dumb-bell configuration the double bond can easily be opened, thus forming the ionized self-interstitial, e.g.

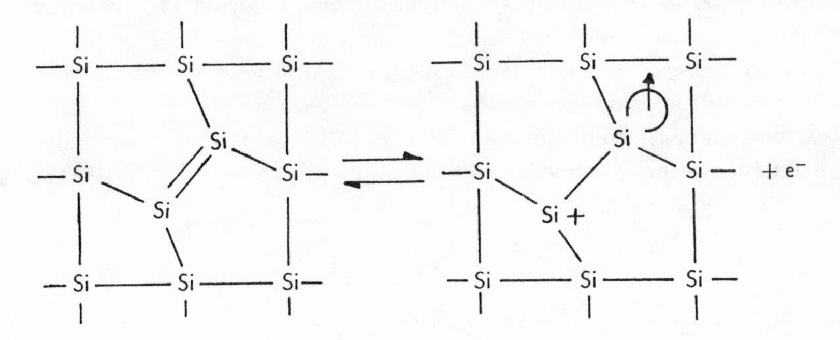

or, in short

$$\mathrm{Si_i} \rightleftharpoons \mathrm{Si_i^+} + \mathrm{e}^- \quad .$$

The self-interstitial is characterized by negative and positive charge states with energy close to the midgap, showing the amphoteric nature of the self-interstitial.

Thermodynamic data for the self-interstitial suffer from the same difficulties as for the vacancy: First of all, the number of allowed sites depends upon the assumed structure of the defect. Second, the self-interstitial seems to have an extended nature. This conclusion is obtained by inspection of Table 3.2, giving the formation entropy and enthalpy of the self-interstitial [3.10].

Table 3.2: Formation entropy and enthalpy of the self-interstitial

T[K]	$\Delta S_\mathrm{i}[k_\mathrm{B}]$	ΔH_i[eV/atom]
570	1	2.6
1320	5.02	2.90
1658	6.11	3.04

The high value of the formation entropy at high temperature suggests that under these conditions the disorder is spread over many silicon atoms. *Taniguchi* et al. [3.11] studied the growth of stacking faults in the temperature range 1100–1200 °C and estimated completely different values (a formation energy of 0.7 eV/atom and a negative formation entropy of $-7.6k_\mathrm{B}$), but paradoxically their results (the large negative formation entropy) lead to the same conclusion, of the extended nature of the self-interstitial. More recently, *Bronner* and *Plummer* [3.12] reconsidered the same problem, confirmed the negative formation entropy (however much larger, $-26k_B$), but obtained a high formation energy. Finally, to conclude this section, we simply state that extended theoretical calculations give a formation enthalpy of around 5 eV but no indications on the formation entropy [3.13].

Still more controversial is the situation concerning the diffusivity of self-interstitials (as well as of vacancies):

Indeed, both non-thermal diffusion at 4 K [3.13] as well as thermally-activated diffusion appreciable only at temperatures of the order of 1000 K [3.11] have been hypothesized.

3.3 Vacancy–Self-Interstitial Pair

We have already observed that when surface reconstruction is impossible, the vacancy and self-interstitial cannot be generated independently of one another, but are formed simultaneously. We shall give what we believe is evidence for this mechanism in Sect. 6.2. We also feel that some discrepancies among different authors concerning vacancies and self-interstitials are actually due to having neglected their generation in pairs. Once generated in pairs, vacancies and self-interstitials can migrate, giving so rise to two free defects, or can remain adjacent, for instance because they are electrostatically stabilized in the ionized forms $v^+ i^-$ or $v^- i^+$. It is not precisely known if such pairs exist or in which ionized state; what seems sure, however, is that a barrier exists against their recombination. For this barrier, both enthalpic [3.14,15] and entropic [3.16] natures have been postulated. The entropic nature is easily understood if one admits that the self-interstitial, and possibly the vacancy, are smeared out over several atomic volumes. According to *Gösele* et al. [3.16], vacancy-interstitial recombination requires the contraction of both defects to about one atomic volume at about the same location. Since the number of microstates associated with extended defects is larger than the number of microstates of the point defect, the contraction implies a decrease of entropy, i.e. an entropy barrier.

Although the v-i pair is presumably stable, its formation enthalpy ΔH_{vi}, which is the sum of the formation enthalpies of a vacancy and a self-interstitial, should be in the range 4–8 eV, according to various estimates of ΔH_i and ΔH_v. The density of pairs which form under equilibrium condition for a process at constant surface is given by [3.17]:

$$N_{vi}^2/(N_v^0 - N_{vi})(N_i^0 - N_{vi}) = \exp(-\Delta H_{vi}/k_B T) \tag{3.3}$$

where all entropic contributions have been ignored. Taking $N_i^0 \simeq N_v^0 = N_{Si}$, for $N_{vi} \ll N_{Si}$, (3.3) is reduced to

$$N_{vi} \simeq N_{Si} \exp(-\Delta H_{vi}/2k_B T).$$

Assuming that $\Delta H_{vi} = \Delta H_v + \Delta H_i$, the concentration of defects formed in a process at constant surface is controlled by a formation enthalpy $(\Delta H_v + \Delta H_i)/2$ and (unless ΔH_v equals ΔH_i) is much lower than the concentration of defects formed in a process at constant volume or pressure, which is controlled by the lowest enthalpy. Note, however that

- vacancy and self-interstitial can exist as a close pair, because of their recombination enthalpy or entropy barrier;
- vacancy and self-interstitial can exist as an ionized pair quite irrespective of the Fermi energy, because of their amphoteric natures;
- according to the value of the Fermi energy, different ionized pairs ($v^+ i^-$ or $v^- i^+$) can be obtained;
- with respect to a pair formed by infinitely separated ionized vacancy and self-interstitial, the close pair requires a lower formation energy, the difference being equal to the Coulomb attractive energy in the absence of lattice reconstruction;
- if in the close pair the v-i distance is less than one lattice distance, the dielectric constant screening the Coulomb interaction may be assumed to be the vacuum one, and this gives an attractive energy of the order of 4 eV;
- the vacancy or the self-interstitial can leave the close pair if replaced by an ionized impurity (e.g., v^- can leave the $v^- i^+$ pair by interaction with an ionized acceptor, such as B^-).

In conclusion, the close ionized v-i pair may be considered as a native defect with a modest formation enthalpy ($\frac{1}{2}\Delta H_{vi} = 2 - 4$ eV) and can be formed everywhere in the crystal irrespective of the possibility of surface reconstruction.

3.4 Stacking Faults

Vacancies and self-interstitials can be present even in non equilibrium concentrations. For instance, an excess of both can be obtained by quenching a sample after high temperature annealing; an interstitial excess and a vacancy deficiency within a few diffusion lengths from the $Si\text{-}SiO_2$ interface can be obtained during silicon oxidation at temperature below 1200 °C. When these defects are in large excess with respect to their equilibrium concentration, they can precipitate in metastable phases.

Extrinsic Stacking Faults Self-interstitial precipitates in discs lying on (111) planes are referred to as *extrinsic stacking faults* (ESFs). Precipitation in (111) planes is a probable process (however not the unique — (113) defects are also known [3.18,19]) because the energy excess is quite small, about 50–60 erg/cm^2 [3.20,21]; this small energy difference comes from the third nearest neighbour arrangement, so that the system is only slightly disturbed compared to other bond-breaking defects, such as dislocations [3.22].

Figure 3.1 shows a scanning electron microscope (SEM) image of the etch pattern after Secco etching of an ESF due to a self-interstitial excess obtained during an oxygen precipitation process.

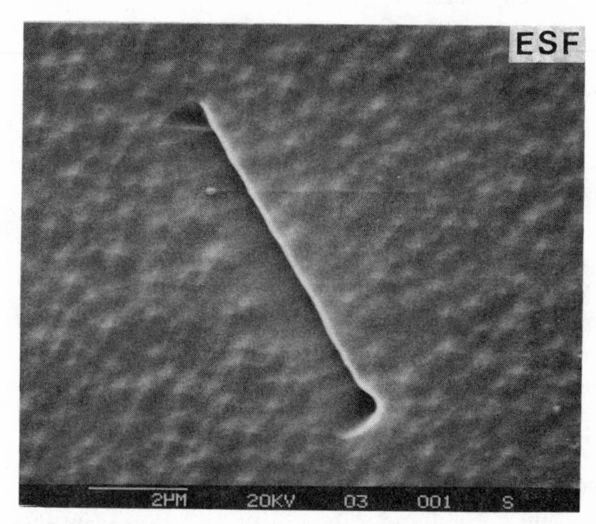

Fig. 3.1 SEM view of an ESF etch pattern after Secco etching

Secco etch is a concentrated HF + $Na_2Cr_2O_7$ aqueous solution which preferentially etches highly stressed regions thus giving evidence for extended defects [3.23].

Precipitation into stacking faults requires a self-interstitial excess and is greatly favoured by the existence of suitable nuclei. These nuclei are often at the surface and several techniques have been proposed for reducing their number (see Sect. 9.1).

Intrinsic Stacking Faults The *intrinsic stacking fault* (ISF) can be seen as a precipitate of vacancies or, equivalently, as a missing disc of silicon atoms in a (111) plane. Though its energy excess should be of the same order as that of the ESF, in a recent review Zulehner and Huber claimed that [3.24]

> "up to now [1982] no intrinsic stacking fault that has been formed by vacancy agglomeration has been detected in silicon, only extrinsic ones".

Shortly afterwards, however, *Claeys* et al. gave TEM evidence for an ISF [3.25], while *Cerofolini* and *Polignano* suggested that ISFs and ESFs can be formed simultaneously during oxygen precipitation, and supported this hypothesis with SEM evidence (Fig. 3.2) [3.26].

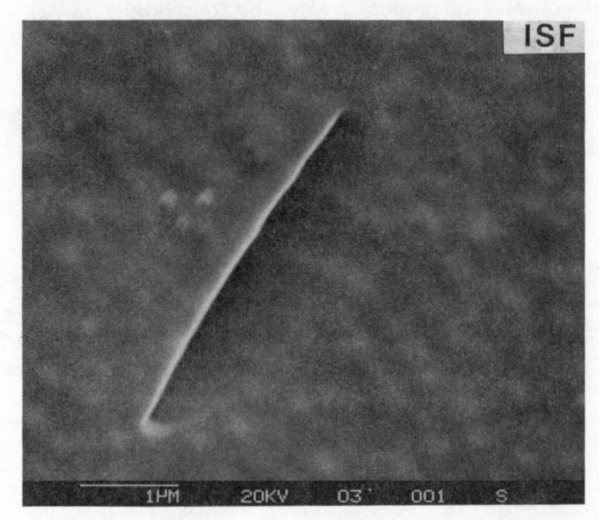

Fig. 3.2 SEM view of an ISF etch pattern after Secco etching

It is likely that, ISFs are also formed by annealing of implanted layers close to the surface, where a vacancy excess is created because of recoil effects [3.27]; Fig. 3.3 gives HREM evidence for an ISF.

Fig. 3.3 HREM view of an ISF

Interestingly enough, ISFs have been observed mainly in situations where oxygen plays an important role. It is possible that the ISF can exist only in these situations.

First-principles calculations of the energy excess of stacking faults give 26 erg/cm^2 for the ESF and 40 erg/cm^2 for the ISF, calculated on the assumption that the ISF actually does not involve broken bonds [3.22]. These calculations suggest that ISFs should also form easily; the rarity of ISF observations indicates that some of the hypotheses used in calculating the formation energy are presumably false.

4. Impurities

While equilibrium defects are formed simply for thermodynamic reasons and can be controlled by prolonged annealing at the desired temperature, impurities (which may be present in silicon even at high concentrations because of the growth technique) cannot usually be removed by simple heat treatments.

Impurities stem either from the ingot preparation process (and in this case are often unwanted), or can be inserted by doping. Table 4.1 lists the major impurities in semiconductor silicon; group III and V elements ('dopants') will be considered extensively in the next section, heavy metals will be considered in Chaps. 7 and 9, and this section will be devoted mainly to oxygen.

Table 4.1: Major impurities in silicon

Elements	Source	Lattice location	Diffusivity
Groups III–V	ingot; doping	substitutional	very slow
Group IV (except C)	doping	substitutional	very slow
carbon	ingot	substitutional	very slow
oxygen	ingot	interstitial	slow
metals	ingot; doping	interstitial	fast

4.1 Impurity Content

The impurity content in silicon depends rather markedly on the growth technique, the commercial single crystal silicon being produced either by the Czochralski method or by the float zone method. Table 4.2 reports the typical oxygen and carbon content for FZ and CZ crystals; for comparison, their solid solubilities at 1200 °C are also reported [4.1]. This comparison shows that the carbon concentration may exceed the solid solubility both in CZ and FZ crystals, while the oxygen concentration may exceed the solid solubility only in CZ crystals.

Table 4.2: Oxygen and carbon content of single crystal silicon

Growth	Oxygen [cm^{-3}]	Carbon [cm^{-3}]
FZ (vacuum)	$< 3 \times 10^{15}$	$< 5 \times 10^{16}$
FZ (argon)	$< 2 \times 10^{16}$	$< 3 \times 10^{17}$
CZ	$4 \times 10^{17} - 2 \times 10^{18}$	$10^{16} - 5 \times 10^{17}$
solid solubility at 1200 °C	5×10^{17}	5×10^{16}

The CZ technique is usually preferred in semiconductor technology because a higher oxygen content is responsible for an increase of the plastic limit τ_p.

Table 4.3 reports the properties of CZ silicon (from updating the [4.2] data); note, however, that different producers have slightly different specifications on impurity content (compare Tables 4.2 and 4.3).

Table 4.3: Properties of Czochralski-grown silicon ingot

Property	Producible range	Commercially available range
diameter[mm]	up to 210	50–155
length[mm]	up to 2200	500–1600
orientations	$\langle 100\rangle, \langle 111\rangle, \langle 110\rangle, \langle 511\rangle$	$\langle 100\rangle, \langle 111\rangle$
dopant concentration[cm^{-3}]		
boron	$10^{14} - 10^{20}$	$2 \times 10^{14} - 10^{20}$
phosphorus	$10^{14} - 6 \times 10^{19}$	$10^{14} - 7 \times 10^{17}$
antimony	$10^{14} - 2 \times 10^{19}$	$10^{18} - 10^{19}$
arsenic	$10^{14} - 8 \times 10^{19}$	$5 \times 10^{18} - 8 \times 10^{19}$
aluminium	$10^{14} - 5 \times 10^{17}$	not standard
gallium	$10^{14} - 10^{18}$	not standard
indium	$10^{14} - 1.5 \times 10^{16}$	not standard
impurity concentration[cm^{-3}]		
carbon	$< 10^{16}$ by selection	$< 5 \times 10^{15} - 5 \times 10^{16}$
oxygen	$2 \times 10^{17} - 2 \times 10^{18}$	$5 \times 10^{17} - 1.4 \times 10^{18}$
crystal defects		
dislocations	free	free
precipitates	free	free

4.2 Oxygen

Oxygen in CZ silicon comes from the reduction of crucible SiO_2 by molten silicon

$$\mathrm{Si} + \mathrm{SiO_2} \rightleftharpoons 2\mathrm{SiO} \uparrow \quad ;$$

SiO, in turn, is incorporated in the melt and hence in the ingot; for more details see [4.2].

Oxygen in silicon holds an interstitial position, bridging two nearest-neighbour silicon atoms:

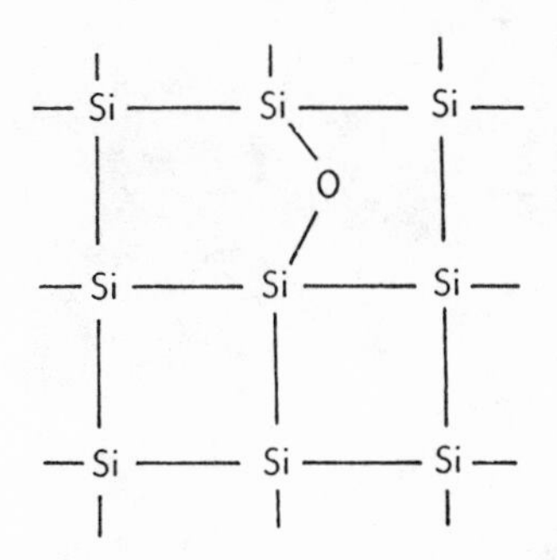

Because of symmetry considerations, 24 equivalent sites around each silicon atom are possible. Although oxygen occupies an interstitial site, its diffusivity is not comparable with that of interstitial metals because of the covalent nature of the Si-O bond. The dependence of the diffusion coefficient on temperature is given by [4.3]

$$D_{\mathrm{O_i}} = 0.16 \exp(-2.53\ \mathrm{eV}/k_{\mathrm{B}}T)\ \ \mathrm{cm^2/s} \quad .$$

Since in CZ crystals the oxygen concentration usually exceeds solid solubility, oxygen precipitation can occur. Precipitation takes place via formation of clusters where a silicon atom is tetrahedrally bound to four oxygen atoms as in SiO_2 (Fig. 8.1 gives a pictorial example); the following scheme represents a two-dimensional configuration:

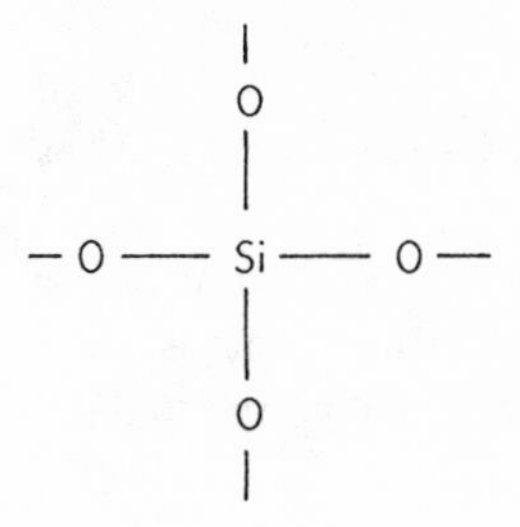

The interface between silicon and precipitated SiO_2 will be characterized by broken, unsaturated or deformed bonds so that it will presumably behave as a generation-recombination centre for electrons and holes.

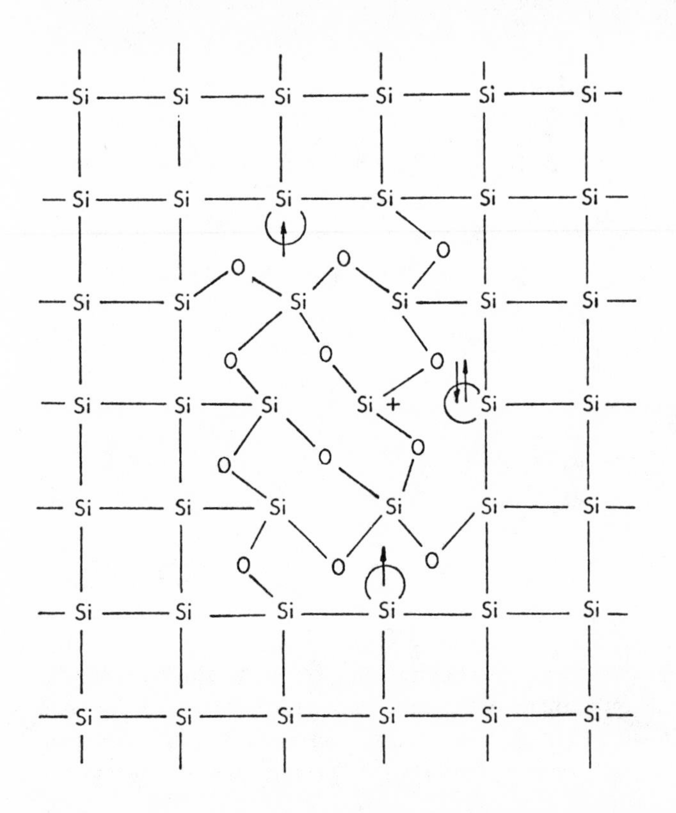

Oxygen precipitation is responsible for the transformation of a silicon volume V_{Si} into a precipitate volume V_{SiO_2}. Since $V_{SiO_2}/V_{Si} \simeq 2$, oxygen precipitation puts the silicon crystal in a compressed state which, if the cluster is large enough, can be relieved by emission of self-interstitials. If precipitated SiO_2 is in the same compressed state as thermally grown SiO_2, the number x of self-interstitials injected per precipitated molecule is of the order of unity ($x \approx 1$) :

$$(1+x)\,\mathrm{Si} + 2\,\mathrm{O_i} \rightleftharpoons \mathrm{SiO_2}\downarrow + x\,\mathrm{Si_i} \qquad x \approx 1 \quad . \tag{4.1}$$

This view [4.4] is perhaps over-simplified [4.5] and will be discussed further in Sect. 6.2.

The kinetics of oxygen precipitation are very complex and may occur homogeneously or heterogeneously on pre-existing nucleation centres. For instance, *Hu* [4.6] suggested that vacancy clusters may be active nucleation centres for oxygen precipitation. This conclusion was derived by observing that oxygen precipitation is retarded by annnealing in an oxygen atmosphere compared with an anneal in nitrogen atmosphere. This can be interpreted by assuming that the interstitial excess generated during oxidation diffuses into the bulk and annihilates the vacancy clusters. Irrespective of the detailed mechanisms through which oxygen precitation occurs, this phenomenon can take place at temperatures high enough to allow oxygen diffusion, but low enough to permit over-saturation. Since the solid solubility increases with temperature, in practice, oxygen precipitation does not occur for processes at temperature above 1250°C.

For lower temperatures, in the range 800 – 1200°C of interest for semiconductor device processing, the occurrence or otherwise of precipitation depends on the interstitial oxygen concentration and silicon thermal history (responsible for precipitation nuclei). More details on the control of oxygen precipitation will be given in Sect. 9.2.

Interstitial oxygen at temperatures around 450°C or 850°C produces complexes known as *thermal donor* and *new thermal donor*, respectively. Extended data on kinetics of formation/destruction of thermal donors and new thermal donors are reported in [4.7].

4.2.1 Thermal Donors

The thermal donor is a defect formed by heat treatments around 450°C and destroyed at temperatures above 550°C. The thermal donor is schematically characterized by two donor energy levels, at about 70 meV and 150 meV from the conduction band, and almost certainly involves oxygen, as its kinetic features indicate:
– the initial formation rate varies as $N_{O_i}^4$;
– the maximum donor concentration varies as $N_{O_i}^3$; and
– the activation energy for the direct reaction equals that for the reverse reaction.
Several models have been proposed to explain phenomena involving thermal donors. For example, and mainly to give an idea of the different arguments which have been advocated to explain the nature of thermal donors, we briefly mention three models often referred to in the literature:

The oldest one, due to *Kaiser* et al. [4.8], explains all kinetic behaviour irrespective of the microscopic nature of the centre. This model assumes that:
i) the formation of Si_mO_n complexes (m, n integers) occurs through consecutive addition of interstitial oxygen atoms: $Si_mO_n + O_i \longrightarrow Si_mO_{n+1}$;
ii) the thermal donor requires the addition of four interstitial oxygen atoms;
iii) the addition of more oxygen atoms destroys the thermal donor.
In the model of Kaiser et al. the thermal donor is a cluster with a well defined, though unspecified, chemical composition $Si_mO_{\overline{n}}$ ($\overline{n} \geq 4$).

A suggestion about the nature of this centre was made by *Oehrlein* and *Corbett* [4.3]: Complexes such as Si_3O_2, Si_4O_3 and Si_5O_4 are assumed to be present after a certain annealing at 450°C. If the complex is not large, it stresses the crystal with an energy, which, however, is insufficient to generate a self-interstitial. Thus both silicon and cluster are compressed, and one can presume that in some conditions the strain is large enough to severely distort Si–O–Si bond angles thus causing a superposition of two lone 2p oxygen orbitals. The two interacting lone pair orbitals result in bonding σ and antibonding σ^* molecular orbitals, while non-interacting 2p orbitals presumably remain unchanged. It suffices therefore to assume that the σ^* antibonding orbitals are in the conduction band to explain the double donor behaviour of the thermal donor:

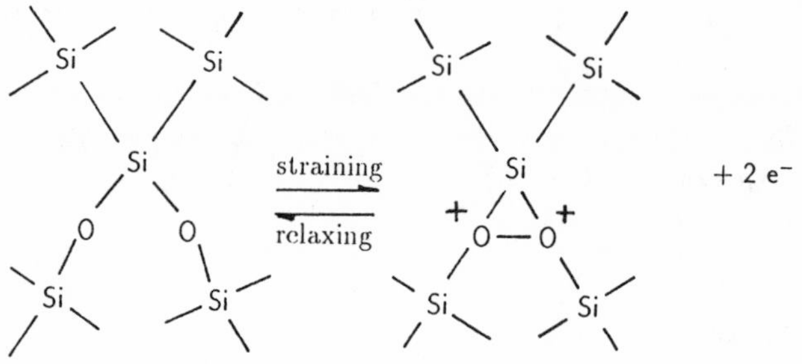

Si Si Si O O Si Si $\underset{\text{relaxing}}{\overset{\text{straining}}{\rightleftharpoons}}$ Si Si Si $^{+}$O—O^{+} Si Si $+ 2e^-$

Larger clusters have an elastic energy sufficiently high to inject self-interstitials into the crystal, thus relieving the stress and inactivating the centre.

The model of Kaiser et al. was criticized by *Helmreich* and *Sirtl* [4.9] on the grounds that the growth rate of the cluster, calculated from the diffusion coefficient of interstitial oxygen at 450 °C, is too low compared to thermal donor generation rate. To avoid this difficulty, Helmreich and Sirtl hypothesized a fast-diffusing form of interstitial oxygen, paired with a vacancy, and assumed the v-O_i pair to be the thermal donor. At present, this model is still considered [4.10], though quantum mechanical calculations indicate that the v-O_i centre should be an acceptor, not a donor [4.11].

Another model for the thermal donor was proposed by *Gösele* and *Tan* [4.12] in order to explain also the high oxygen diffusivity at 450 °C. This model is based on the following assumptions:

i) interstitial oxygen atoms react to give a gas-like O_2 molecule embedded in the crystal:

$$O_i + O_i \rightleftharpoons O_2 \quad ;$$

ii) two oxygen molecules react to give a molecular complex O_4:

$$O_2 + O_2 \rightleftharpoons O_4 \quad ;$$

iii) the complex O_4 is the thermal donor; and
iv) the addition of O_i to O_4 inactivates the complex.

The above description does not exhaust all the proposed models of thermal donors; a recent, perhaps complete, review was given by *Bourret* [4.13].

4.2.2 The New Thermal Donor

The new thermal donor was discovered only recently [4.14]. It is obtained after prolonged heating in the range 550 – 900°C and is not destroyed by heating up to 1000°C [4.7]. This centre is almost certainly related to both oxygen and carbon, but no model has yet been accepted for it.

Table 4.4 compares the properties of the thermal and new thermal donors [4.15].

Table 4.4: Characteristics of oxygen-related donors

Property	Thermal donor	New thermal donor
formation temperature	400–550 °C	550–900 °C
donor killer	heat treatments at 600 – 800 °C	?
generation rate	fast	slow
oxygen concentration	high ⇒ fast	high ⇒ fast
carbon concentration	high ⇒ slow	high ⇒ fast
pre-heating		450 °C ⇒ fast
energy level	70 meV, 150 meV	17 meV
model	$Si_mO_4 (m \simeq 2-3)$; v-O; O_4	?

4.3 Oxygen Precipitates

Oxygen precipitates in silicon have been studied extensively by transmission electron microscopy. It now seems well established that SiO_2 precipitates are always in an amorphous form [4.16], thus disproving the previous suggestion that SiO_2 can precipitate as crystalline coesite [4.17]. Depending on the formation temperature, silicon reacts differently to oxygen precipitation:

1) at moderate temperature, but for relatively large precipitates, the stress energy associated with the increase of volume involved in the $Si+2O_i \rightarrow SiO_2$ reaction is sufficient to form self-interstitials or v-i pairs, and ESFs (or possibly ISFs) may form.

2) At higher temperature the plastic limit of silicon is lowered and the stress energy associated with large SiO_2 precipates is absorbed by the formation of dislocations.
3) At still higher temperature the crystal can rearrange and the reconstruction of the $(SiO_2)_{prec}$-Si interface is possible. This reconstruction leads to the formation of the octahedral equilibrium shape of silicon, and therefore the morphology of SiO_2 is of octahedral amorphous islands embedded in the crystalline matrix.

TEM photographs showing the occurrence of all these possibilities have been reported by *Shimura* and *Craven* [4.18]. Hence, although the scheme of Table 4.5 does not exhaust all phenomena concerning oxygen, it seems however a reasonable rationalization of its actual evolution:

Table 4.5: Oxygen evolution after heat treatments at moderate and high temperature

Temperature	Defect	Structure
room temperature	interstitial oxygen	Si_2O
450°C	coherent precipitates	$Si_mO_2, Si_mO_3, \underline{Si_mO_4}$, increasing stress $\longrightarrow$
550°C	incoherent precipitates	Si_mO_5,... stress is relieved by i or v-i injection
950°C	precipitates and SFs	$(SiO_2)_{prec}$
1000°C	precipitates and dislocations	$(SiO_2)_{prec}$
1100°C	octahedral silica	SiO_2

In this scheme $m \simeq 2 - 3$, the assumed thermal donor is underlined, and the word 'precipitate' does not necessarily mean a full stoichiometric SiO_2 compound.

5. Dopants

Group III and group V impurities play a fundamental role in the electronic, equilibrium and transport properties of silicon and are usually referred to as *dopants.* Dopants are defined as *acceptors* or *donors* when they belong to group III or V, respectively. The properties of doped silicon are usually considered in the framework of a relatively simple theory — the *standard theory of shallow dopants* (STSD).

5.1 The Standard Theory

5.1.1 Electronic Properties

In the context of effective mass approximation (EMA) the following assumptions are made:

> EMA1: the dopant is tetrahedrally bound to four neighbouring silicon atoms;
>
> EMA2: the dopant is charged (positively if donor, negatively if acceptor) because the impurity is not isovalent with silicon;
>
> EMA3: the dopant charge, due to the tetrahedral configuration, is the source for a Coulomb potential which superimposes on the atomic potential generated by the distribution of valence and core electrons and by the nucleus; and
>
> EMA4: the interaction of a quasi-particle in the conduction or valence band with a dopant does not modify the quasi-particle (i.e., quasi-particle properties such as the effective mass) so that the interaction is uniquely described in terms of the dopant potential.

When conditions EMA1 to EMA4 are satisfied, the dopant is said to be *shallow.* With certain changes, the EMA can also be applied to interstitial donors, such as alkaline ions, where the charge state of the impurity can be regarded as given. In STSD, the quasi-particle is described by its unperturbed lattice properties.

The band structure of silicon is characterized by a band gap $E_g = 1.17$ eV at 0 K, with minima in the conduction band at values of quasi-momentum $\mathbf{k}$ directed along the $\langle 100 \rangle$ directions, and valence band maxima at $\mathbf{k} = \mathbf{0}$ [5.1]. Because of symmetry considerations, six equivalent valleys define the lowest lying states in conduction band; each valley is described by an anisotropic mass tensor m^*_{ik} $(i, k = x, y, z)$ with an ellipsoidal symmetry, the longitudinal effective mass being $m^*_{\mathrm{el}} = 0.97\ m_e$ and the transverse effective mass being $m^*_{\mathrm{et}} = 0.19\ m_e$, where m_e is the electron mass. The density-of-states effective mass, which takes into account the six valleys, is given by

$$(m^*_{\mathrm{n}})^{\frac{3}{2}} = 6(m^*_{\mathrm{el}} m^{*2}_{\mathrm{et}})^{\frac{1}{2}} ,$$

i.e., $m^*_{\mathrm{n}} = 1.1\ m_e$.
The valence band is described in terms of three types of hole:
– the *heavy hole*, with effective mass $m^*_{\mathrm{hh}} = 0.52\ m_e$,
– the *light hole*, with effective mass $m^*_{\mathrm{lh}} = 0.16\ m_e$,
both with dispersion relationships with maxima at the top of the valence band, and
– the *split-off hole*, with effective mass $m^*_{\mathrm{soh}} = 0.16\ m_e$
and with maximum at 45 meV below the valence-band edge. For many considerations the split-off hole can be ignored and the four-fold degenerate dispersion relationship at quasi-momentum $\mathbf{k} = \mathbf{0}$ can be regarded as a unique quasi-particle with scalar effective mass

$$m^*_{\mathrm{p}} = (m^{*\frac{3}{2}}_{\mathrm{hh}} + m^{*\frac{3}{2}}_{\mathrm{lh}})^{\frac{2}{3}} = 0.58\ m_e$$

and spin $\frac{3}{2}$ [5.2].

In STSD a shallow dopant produces a Coulomb field screened by the silicon dielectric constant $\varepsilon_{\mathrm{Si}}$, the potential at a distance $\mathbf{r}$ from the dopant nucleus being

$$V_{\mathrm{coul}}(r) = \frac{1}{4\pi\varepsilon_0\varepsilon_{\mathrm{Si}}} \frac{ze}{|\,\mathbf{r}\,|} \tag{5.1}$$

where ε_0 is the vacuum dielectric constant, e is the positive unit charge, and z is the net dopant charge ($z = 1$ for donors, $z = -1$ for acceptors). In addition to this potential, one must also consider the potential generated by the distribution of core electrons. However, if the guest dopant which replaces the host is adjacent to it in the Periodic Table (i.e., aluminium and phosphorus in silicon, gallium and arsenic in germanium) the electron distribution of host and guest atoms are similar and the potential (5.1) can be regarded as adequate to describe the interaction between dopant and quasi-particle.

The Hydrogenic Model The analysis of the interaction is particularly simple if the effective mass m^*_{ik} is a scalar quantity m^*. In this case the eigenvalue problem for the quasi-particle is easily solved by scaling the result for the hydrogen atom, i.e. by replacing the m_e by m^* and ε_0 by $\varepsilon_0\varepsilon_{\rm Si}$.

This result follows from an accurate analysis of the problem, which shows that the *electronic* problem (requiring the solution of the single electron Schrödinger equation in the Coulomb potential superimposed on the crystal periodic potential) can be reduced to the Schrödinger equation for the *quasi-particle* in the Coulomb potential alone [5.3–6]. In particular, the ionization energy $E^*_{\rm ion}$ and the radius a^* of the ground state orbital can be expressed in terms of the ionization energy of the hydrogen atom $E^{\rm H}_{\rm ion}$ ($E^{\rm H}_{\rm ion} = 13.6$ eV) and Bohr radius $a^{\rm H}_0$ ($a^{\rm H}_0 = 0.53$ Å) as

$$E^*_{\rm ion} = \left(\frac{1}{\varepsilon_{\rm Si}}\right)^2 \frac{m^*}{m_{\rm e}} E^{\rm H}_{\rm ion} \quad , \tag{5.2}$$

$$a^* = \varepsilon_{\rm Si} \frac{m_{\rm e}}{m^*} a^{\rm H}_0 \quad . \tag{5.3}$$

To estimate the orders of magnitudes, taking $m^* = m_e$ from (5.2) and (5.3) we have $E^*_{\rm ion} \simeq 100$ meV and $a^* \simeq 6$ Å. It is obvious that the treatment is self-consistent only if a^* greatly exceeds the atomic radius of the impurity and thus does not feel the details of the atomic potential of the dopant.

Consider now the ionization energy $E^0_{\rm ion}$ of the hydrogen-like atom formed by the actual quasi-particle; its value is determined by the complexity of the band structure (resulting in an anisotropic tensorial effective mass for the electron and in a spin-$\frac{3}{2}$ particle for the hole), and even large deviations of $E^0_{\rm ion}$ from $E^*_{\rm ion}$ are expected to occur. Sections 5.2 and 5.3 will consider the problem of the real value of $E^0_{\rm ion}$, which is a lattice property independent of dopant nature.

Central-Cell Correction The complexity of the band structure is not the only factor which may influence the ionization energy; the core electron distribution may be responsible for large deviations of the experimental ionization energy of a given dopant, $E_{\rm ion}$, from $E^0_{\rm ion}$. The difference $E_{\rm ion} - E^0_{\rm ion}$ will be defined as the *chemical shift* ΔE of the dopant.

In general, the EMA assumes a potential of the type

$$V(\mathbf{r}) = V_{\rm loc}(\mathbf{r}) + V_{\rm coul}(r) \tag{5.4}$$

where $V_{\rm loc}(\mathbf{r})$ is the potential due to the electron distribution and responsible for the deviations of $E_{\rm ion}$ from $E^0_{\rm ion}$. The *electron* eigenvalue problem for a potential of type (5.4) can still be considered perturbatively and the solution is still expressed in terms of band properties, provided that $V_{\mathrm{loc}}(\mathbf{r})$ is confined within the first cell. In such a case one speaks of a central-cell correction (CCC) [5.6,7].

Irrespective of the details of the solution, we observe that if the potential $V_{\mathrm{loc}}(\mathbf{r})$ is attractive, it may influence even considerably the ionization energy, while if it is repulsive, then the correction to the ionization energy is small (in fact, a repulsive potential tends to localize the particle beyond the first cell, where by hypothesis $V_{\mathrm{loc}}(\mathbf{r}) = 0$); values of E_{ion} much lower than E^{0}_{ion} are therefore incompatible with the CCC-EMA. In other words, large negative values of chemical shift cannot be explained by the EMA in the CCC.

5.1.2 Equilibrium Properties

Hypotheses EMA1 to EMA4 are useful for determining the electronic properties of doped silicon. The STSD, however, is not exhausted at this stage but aims at describing equilibrium and transport properties too. To do this, STSD requires additional assumptions. For equilibrium properties these are:

> E1: The electron and hole densities, n and p, can be varied by changing the dopant concentration. However, their product np is a constant at constant temperature

$$np = n_{\mathrm{i}}^{2} \quad . \tag{5.5}$$

In (5.5) n_{i} is the intrinsic concentration of electrons and holes, i.e. the concentration in the absence of any dopant,

$$n_{\mathrm{i}} = (N_{\mathrm{C}}N_{\mathrm{V}})^{\frac{1}{2}} \exp(-E_{\mathrm{g}}/2k_{\mathrm{B}}T) \quad .$$

N_{C} and N_{V} are the density of states in the conduction and valence bands, respectively:

$$N_{\mathrm{C}} = 2\left(\frac{2\pi m_{\mathrm{e}}k_{\mathrm{B}}T}{h^2}\right)^{\frac{3}{2}}\left(\frac{m^*_{\mathrm{n}}}{m_{\mathrm{e}}}\right)^{\frac{3}{2}} \quad ,$$

$$N_{\mathrm{V}} = 2\left(\frac{2\pi m_{\mathrm{e}}k_{\mathrm{B}}T}{h^2}\right)^{\frac{3}{2}}\left(\frac{m^*_{\mathrm{p}}}{m_{\mathrm{e}}}\right)^{\frac{3}{2}} \quad ,$$

and h is Planck's constant.

> E2: The ionization of neutral donors and acceptors is an equilibrium process

$$A^- \cdots h^+ \rightleftharpoons A^- + h^+$$

$$D^+ \cdots e^- \rightleftharpoons D^+ + e^- \ .$$

The equilibrium constants of these reactions are given by

$$K_\mathrm{D} = \frac{n\,N_{\mathrm{D}^+}}{N_{\mathrm{D}^0}} = \frac{n\,N_{\mathrm{D}^+}}{N_\mathrm{D} - N_{\mathrm{D}^+}} = \frac{N_\mathrm{C}}{g}\exp(-\frac{E_\mathrm{ion}}{k_\mathrm{B}T}) \ , \tag{5.6}$$

$$K_\mathrm{A} = \frac{p\,N_{\mathrm{A}^-}}{N_{\mathrm{A}^0}} = \frac{p\,N_{\mathrm{A}^-}}{N_\mathrm{A} - N_{\mathrm{A}^-}} = \frac{N_\mathrm{V}}{g}\exp(-\frac{E_\mathrm{ion}}{k_\mathrm{B}T}) \ , \tag{5.7}$$

where g is the degeneracy factor of the fundamental state ($g = 2$ for electrons because of spin-$\frac{1}{2}$ degeneracy, $g = 4$ for holes because of spin-$\frac{3}{2}$ degeneracy) and N_{D^z} (N_{A^z}) is the donor (acceptor) concentration in the ionization state with charge z. Equations (5.6) and (5.7) are derived assuming the Boltzmann statistics and fail when applied to degenerate semiconductors.

E3: The condition of local electroneutrality,

$$N_{\mathrm{D}^+} - N_{\mathrm{A}^-} + p - n = 0 \ , \tag{5.8}$$

holds true everywhere.

If N_D and N_A are known, we have four equations (5.5-8) in the four unknowns $n, p, N_{\mathrm{D}^+}, N_{\mathrm{A}^-}$, which can be solved, though not in a closed form. Therefore, assumptions E1 to E3 allow the electron and hole concentrations to be calculated.

The above equations show that, for given N_D and N_A, electron and hole concentrations depend on the nature of the dopant only through their dissociation constant, i.e., through their ionization energy E_ion. It must be noted that the model works only for dilute solutions; in the high density limit E_ion varies and eventually vanishes as N_D (and N_A) are increased. This phenomenon can be interpreted in terms of semiconductor-metal transition; the n-type doped silicon [1] undergoes this transition for concentrations in the range $10^{18} - 10^{19}\ \mathrm{cm}^{-3}$ [5.8].

It is interesting to remark that the equilibrium properties of electrons and holes are calculated in a way perfectly similar to the one used to calculate the dissociation properties of weak acids and bases in a diluted aqueous

[1] Doped silicon is defined as p-type or n-type according to whether the hole or electron concentration predominates. The carriers with higher concentration are referred to as majority carriers, the ones with lower concentration as minority carriers. A p- (n-) type region, doped at concentration high enough to allow the semiconductor-to-metal transition, will be referred to as p^+ (n^+).

solution. This analogy goes beyond the calculation of the ionized fraction, and the whole Debye–Huckel theory of ionic solutions [5.9] can be restated for moderately doped semiconductors. The major difference is that dopants are considered immobile, i.e., N_{D} and N_{A} are given functions of position $\mathbf{x}$; this difference plays an important role in electron device physics, where steep profiles often occur, but can be ignored in most physical-chemical situations. This similarity was noted in early works too [5.10], and can be extended to establish an analogy between the electrolytic cell and the p-n junction.

5.1.3 Transport Properties

The motion of a quasi-particle in an external field is characterized by two regimes: *ballistic* and *dissipative*.

In the ballistic regime the carrier moves under the action of the external force $\mathbf{F}$ alone. Its group velocity $\mathbf{v}_{\mathrm{g}}$ is described in terms of Newton's second law:

$$\dot{\mathbf{v}}_{\mathrm{g}} = \frac{1}{m_{ik}}\mathbf{F} \tag{5.9}$$

provided that $\mathbf{F}$ satisfies suitable conditions (slow variation). In the absence of any scattering mechanism, under ballistic motion the quasi-particle can in principle reach any velocity. Does a mechanism exist to limit the velocity at most to c, the velocity of light, and is it internal to the theory itself? The answer is affirmative, as the maximum allowed group velocity w is given by the maximum of $(1/\hbar) \mid \mathrm{d}E/\mathrm{d}\mathbf{k} \mid$, where $E(\mathbf{k})$ is the dispersion relationship in the band. Thus the limiting velocity w is a band property. To give an order of magnitude, for electrons in the conduction band $w \simeq 6 \times 10^7$ cm/s [5.11].

In practice, however, high-field experiments show that the maximum drift velocity is limited to a value v_{sat} around 10^7 cm/s, quite irrespective of T in the range $4 - 300$ K [5.12]. This fact is interpreted by assuming that the drift velocity is limited by other mechanisms not considered by the band description, i.e. scattering by phonons, and by ionized and neutral dopants.

When these mechanisms are present, the motion can no longer be viewed as ballistic, but becomes dissipative. In this regime, after a transitory period the drift velocity $\mathbf{v}$ becomes proportional to the external field, provided that the field is weak enough.

A particularly interesting external field is the electric one $\mathbf{E}_{\mathrm{el}}$; in this case, the constant of proportionality, μ, linking $\mathbf{v}$ to $\mathbf{E}_{\mathrm{el}}$,

$$\mathbf{v} = \mu \mathbf{E}_{\mathrm{el}} \quad ,$$

is known as mobility.

In STSD the mobility is calculated with the following hypotheses [5.13]:

T1: the factor which limits the free motion of quasi-particles is scattering by: 1) phonons, 2) ionized shallow dopants, and 3) neutral shallow dopants;

T2: these events are independent of one another; and

T3: their statistical features are described by the Boltzmann transport equation in the relaxation-time approximation.

Under assumptions T1 to T3 the bulk mobility can be written as the harmonic composition of the lattice mobility μ_l (free motion limited by scattering from phonons, independent of dopant concentration), the ionized impurity mobility μ_i (Rutherford scattering from ionized impurities), and the neutral impurity mobility μ_n (scattering from neutral shallow dopants),

$$\frac{1}{\mu} = \frac{1}{\mu_l} + \frac{1}{\mu_i} + \frac{1}{\mu_n} \quad .$$

Models for the calculation of μ_l, μ_i, and μ_n from first principles (band structure, scattering mechanisms and Boltzmann equation) are known [5.13] and accurately fit the experimental data for Si:B, Si:P and Si:As for dopant concentration below 10^{18} cm^{-3} in the temperature range 100 – 300 K [5.14]. Experimental data for Si:In are much less accurately described by current theories [5.15].

The room temperature dependence of the mobility μ on carrier concentration n in Si:P and Si:As is accurately described by the following seven-parameter expression [5.16]

$$\mu = \mu_0 + \frac{\mu_{max} - \mu_0}{1 + (n/C_r)^\alpha} + \frac{\mu_1}{1 + (C_s/n)^\beta} \quad ,$$

where $\mu_{max}, \mu_0, \mu_1, C_r, C_s, \alpha,$ and β are empirical parameters.
In Si:B samples the mobility μ depends on carrier concentration p as

$$\mu = \mu_0 \exp(-p_c/p) + \frac{\mu_{max}}{1 + (p/C_r)^\alpha} + \frac{\mu_1}{1 + (C_s/p)^\beta}$$

which, compared with the expression for the mobility in n-doped silicon, requires one additional empirical parameter, p_c. The eight parameters μ_{max}, μ_0, μ_1, C_r, C_s, α, β, and p_c are listed in Table 5.1.
Additional hypotheses, not treated here, are required to describe surface mobility, mobility in inversion layers, in polycrystalline silicon, or in the metallic regime.

Table 5.1: Mobility parameters in Si:As, Si:P and Si:B

Parameter	Units	Arsenic	Phosphorus	Boron
μ_0	$cm^2V^{-1}s^{-1}$	52.2	68.5	44.9
μ_{max}	$cm^2V^{-1}s^{-1}$	1417	1414	470.5
μ_1	$cm^2V^{-1}s^{-1}$	43.4	56.1	29.0
C_r	cm^{-3}	9.68×10^{16}	9.20×10^{16}	2.23×10^{17}
C_s	cm^{-3}	3.43×10^{20}	3.41×10^{20}	6.10×10^{20}
α		0.680	0.711	0.719
β		2.00	1.98	2.00
p_c	cm^{-3}	—	—	9.23×10^{16}

5.2 Group V Donors

Table 5.2 presents the ionization energies of group V donors in silicon as given by *Milnes* [5.17].

Table 5.2: Experimental ionization energy of group V impurities in silicon

Element	E_{ion}[meV]
N	140
P	44
As	49
Sb	39
Bi	69

The theoretical calculation of E^0_{ion} for electrons is a difficult task because of the tensorial nature of the effective mass in the conduction band. A way to deal with this problem is described for instance in [5.6]. Here it suffices to note that if we describe the conduction electron by a scalar effective mass $m^*_e = (m^*_{el} m^{*2}_{et})^{\frac{1}{3}} = 0.32\ m_e$, we have an ionization energy $E^*_{ion} \simeq 30$ meV, while taking the density-of-state effective mass $m^*_n = 1.1\ m_e$ we have an ionization energy of about 100 meV. An inspection of Table 5.2 shows that, neglecting nitrogen, the other donors have ionization energies around 50 meV; this value can therefore be assumed as representative of E^0_{ion}. This fact can also be interpreted in the following way: the electron anisotropy can simply be taken into account by defining another scalar effective mass m^{**}_e according to

$$\frac{m^{**}_e}{m_e} = \varepsilon^2_{Si} \frac{E^0_{ion}}{E^H_{ion}}$$

instead of m_e^*. In this way, the anisotropy of the quasi-particle is reduced to an effective-mass renormalization.

The differences $E_{ion} - E_{ion}^0$ for all donors except nitrogen can be explained in terms of small chemical shifts [5.18]. For nitrogen the situation is more complex because the perturbation (the chemical shift) is larger than the unperturbed value (E_{ion}^0). It is therefore possible that in this case some of the basic assumptions of the EMA do not hold true; for instance, it can be hypothesized that the neutral state of nitrogen is of the type

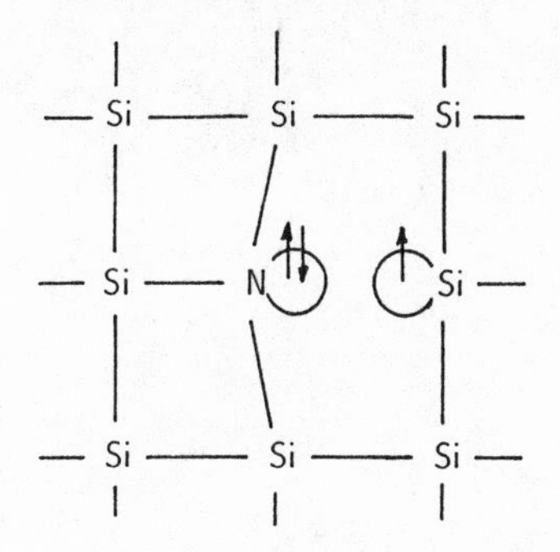

(the electronic configuration of nitrogen being the same as in ammonia). Ionization should then involve a change of electronic configuration:

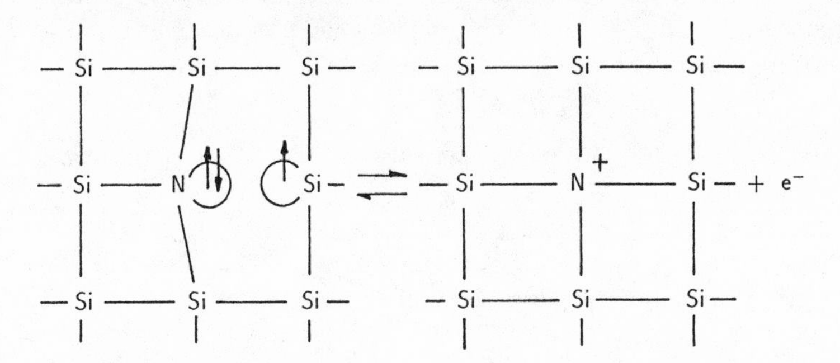

5.3 Group III Acceptors

We have just seen how group V impurities except nitrogen can be considered as shallow donors, adequately described by a *scalar* EMA, provided that the electron effective mass is renormalized. The renormalized mass, $m_e^{**} = 0.5\ m_e$, explaining the observed ionization energies, is intermediate between the single-valley electron effective mass, $m_e^* = 0.32\ m_e$, and the density-of-states effective mass, $m_n^* = 1.1\ m_e$.

A similar procedure cannot be repeated for the hole (to deal, for instance, with valence-band warpage) because the ionization energies of group III elements vary greatly within the group — from 44 meV for boron to 246 meV for thallium. Table 5.3 collects the data on optical ionization energies of group III acceptors as reported by *Jones* et al. [5.19]. The major conclusions of this section are not influenced by the small discrepancies among different sources on the last significative figure of the ionization energy.

Table 5.3: Ionization energy of group III impurities in silicon

Element	E_{ion}[meV]
B	44
Al	69
Ga	73
In	156
Tl	246

The first explanation suggested for the values of E_{ion} for group III impurities is due to *Morgan* [5.20]. In this idea, we observe, first of all, that an estimate of E^0_{ion} on the grounds of a hole effective mass equal to the density-of-states effective mass ($m_p^* = 0.58m_e$) gives $E^0_{\text{ion}} = 57$ meV. Thus we compare (Table 5.4) the difference $E_{\text{ion}} - E^0_{\text{ion}}$ (tentatively, the chemical shift) of an impurity, with its tetrahedral radius in silicon as given in Wolf's handbook [5.21].

Table 5.4: Ionization energy, chemical shift and tetrahedral radius of group III acceptors in silicon

Element	E_{ion}[meV]	$(E_{\text{ion}} - E^0_{\text{ion}})$[meV]	r[Å]	$(r - r_{\text{Si}})$[Å]
B	44	− 13	0.88	− 0.29
Al	69	12	1.26	0.09
Ga	73	16	1.26	0.09
In	156	99	1.44	0.27
Tl	246	189	1.47	0.30

This comparison shows that the difference $\Delta E = E_{\text{ion}} - E^0_{\text{ion}}$ is never negligible and has some interesting behaviour:, 1) ΔE has the same sign as $r - r_{\text{Si}}$; 2) despite the core electron cloud of Al ($Z = 13$) being very similar to that of Si ($Z = 14$), ΔE_{Al} is not negligible; 3) despite the core electron clouds of Al ($Z = 13$) and Ga ($Z = 31$) being very different, $\Delta E_{\text{Al}} \simeq \Delta E_{\text{Ga}}$; 4) the tentative chemical shift increases with $r - r_{\text{Si}}$. These facts show that the different trivalent impurities are distinguished mainly by their tetrahedral radii, and suggest that the chemical shift is associated with the energy required for straining the lattice. If this view is correct, the main contribution to ΔE must be elastic, i.e., must increase roughly in proportion to $(r - r_{\mathrm{Si}})^2$. Figure 5.1 supports this hypothesis.

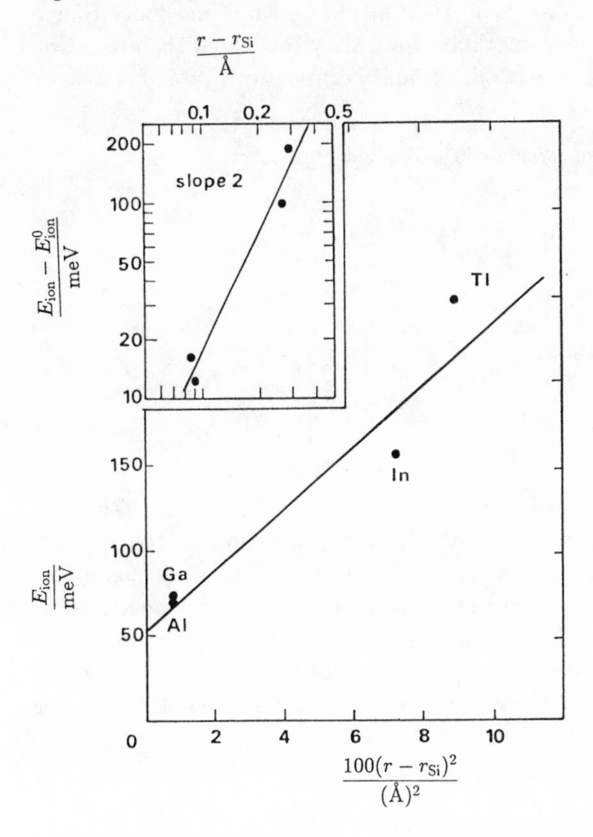

Fig. 5.1 Lin-lin and log-log plots of the ionization energies vs tetrahedral radii for acceptors

However, the idea of their elastic origin does not allow the observed chemical shifts to be explained by STSD. In fact, this problem can be considered in the EMA by taking the deformation potential due to the difference of tetrahedral radii as a CCC. This problem can be solved, and the solution shows that even large differences of tetrahedral radii (up to 0.2 Å) are responsible for only a modest chemical shift, of a few millielectronvolt [5.22]. The problem of explaining the observed chemical shift will be referred to as the 'group III acceptor puzzle'.

5.3.1 Group III Acceptors as Shallow Centres

EMA with CCC The first satisfactory attempt to solve the puzzle mentioned above, is due to *Lipari* et al. [5.23]. These authors considered the postulated correlation $\Delta E \propto (r - r_{Si})^2$ as a 'red herring' and looked for a completely different explanation. First of all, they compared the ionization energies of group III acceptors in silicon and germanium (Table 5.5):

Table 5.5: Ionization energies of acceptors in silicon and germanium

Element	E_{ion}[meV]	
	Silicon	Germanium
B	45.83	10.8
Al	70.42	11.14
Ga	74.16	11.30
In	156.94	11.99
Tl	247.67	13.43

They then added to the Coulomb potential (5.1) (corrected to take into account the frequency dependence of the dielectric function) a strong short-range potential containing only one free parameter (α') which accounts for differences in the electronic structures of the ionic cores of various acceptors as well as differences in the lattice relaxation around them. Then, in view of the inherent difficulties in calculating α' from first principles, they selected the value of this parameter to reproduce the observed ground-state binding energies. Their results are presented in Table 5.6.

Table 5.6: Best fit values of α'

Element	$\alpha'[a_0^H]$	
	Silicon	Germanium
B	3.00	3.00
Al	1.01	1.10
Ga	0.93	0.93
In	0.73	0.70
Tl	0.63	0.57

Inspection of this table shows that for any given acceptor the value of α' in silicon is about the same as in germanium. This fact led Lipari et al. to conclude that "the quasi-transferability of the short-range potentials from Si to Ge [...] clearly indicates that the short-range part of the impurity potential is more related to the atomic properties of the impurity than to the properties of the host lattice".

This conclusion seems to deny the possibility that the chemical shift is due to an elastic effect. In our opinion, however, the calculations by Lipari et al. do not solve the group III acceptor puzzle, but simply shift it to a problem of explaining the correlation between the strength of the short-range potential (through α') and $r - r_{Si}$.

In addition, the STSD also seems unable to explain the following pieces of experimental evidence on chemical, equilibrium and transport properties of acceptors in silicon:

1. Acceptors in silicon are deactivated by atomic hydrogen [5.24-29].
2. An almost complete electrical activation at room temperature up to a concentration of some 10^{18} cm^{-3} (obtainable in a metastable phase) is observed in Si:In [5.30].
3. The ionization energy of aluminium and gallium determined by optical methods, E_{ion}, is a little different from the one determined by thermal methods, E_{th} [5.17]. For indium, moreover, E_{ion} and E_{th} are significantly different [5.31]. Table 5.7 illustrates these differences.

Table 5.7: Ionization energies in silicon determined with optical and thermal methods

Element	E_{ion}[meV]	E_{th}[meV]
B	44	45
Al	69	57
Ga	73	65
In	156	18
Tl	246	?

4. 'Supershallow levels', i.e., levels with thermal ionization energy much lower than the one predicted by the EMA, are observed in neutron-irradiated Si:Ga [5.32,33] after annealing at moderate temperature (in the range 400 – 700°C), and in Si:In samples heated at temperature so high, i.e. above 1100°C, to suggest that they are not associated with process defects, but rather with the intrinsic dopant centre [5.34].
5. A non-linear conductivity increase of several orders of magnitude is observed in illuminated Ge:Ga samples, even for frequency ν of incident radiation much lower than the ionization energy of gallium in germanium, $h\nu \simeq 2.5$ meV against $E_{ion} \simeq 11$ meV [5.35].

To accomodate these facts, and, in particular, items 2 and 3, another attempt to partially solve the group III acceptor puzzle was proposed by *Cappelletti* et al. [5.15], as explained in the following.

Change of Effective Mass In an extended study of the thermodynamic and kinetic properties of the Si:In system, *Cerofolini* et al. [5.30,34,36] found a rather strange behaviour of the indium-doped layer. In particular, they found that indium electrical activation at room temperature remains of the order of unity up to concentrations of 10^{17} cm^{-3} (while STSD predicts that in this concentration range electrical activation should be around 10 %), while the mobility is about one half of the expected value. *Cappelletti* et al. proposed a model to explain this discrepancy [5.15]. Their idea was that the large difference of tetrahedral radii $r_{\mathrm{In}} - r_{\mathrm{Si}}$ is relaxed over a cloud with many atoms. In spite of the deformation, the concept of band remains valid, but the valence band structure (parametrized by the effective mass) varies from point to point. It was then suggested that the local effective mass m_{p}^{**} is the one which explains the experimental ionization energy in the hydrogenic model

$$m_{\mathrm{p}}^{**} = m_{\mathrm{p}}^{*}(E_{\mathrm{ion}}/E_{\mathrm{ion}}^{0}) \ . \qquad (5.10)$$

Since any increase of m_{p}^{**} results in an increase of electrical activation [because $N_{\mathrm{V}} \propto (m_{\mathrm{p}}^{**})^{\frac{3}{2}}$] and a decrease of mobility [because $\mu_{\mathrm{p}} \propto (m_{\mathrm{p}}^{**})^{-\frac{1}{2}}$], eq. (5.10) is perhaps useful to explain the observed activation and mobility data. But although the authors initially claimed that (5.10) *quantitatively* removes the discrepancies, in a later paper *Cappelletti* et al. [5.37] reduced the strength of this statement.

5.3.2 Group III Acceptors as Deep Centres

The basic ideas to explain all the above evidence are the following [5.38,39]:

1. Deep ground state.
   In the ground state, the acceptor atom has a deep sp^2 hybridization, i.e. forms covalent bonds with three nearest-neighbour silicon atoms, the fourth atom remaining with a dangling bond,

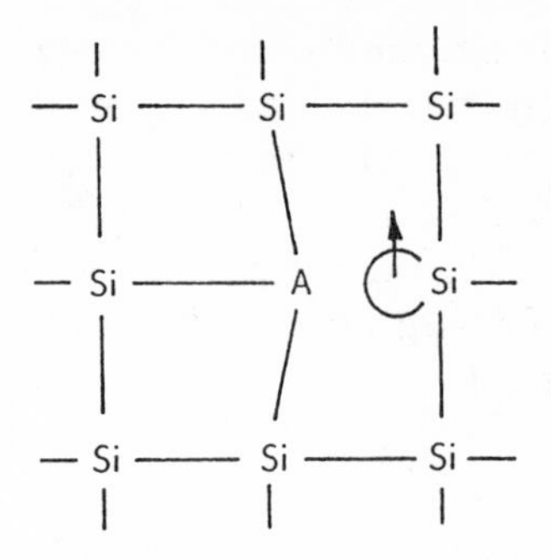

Even in the sp^2 state, a cloud of displaced silicon atoms surrounds the acceptor atom. This cloud is large enough to be considered as a small macro-system embedded in the crystal. The crystal can be thought of as a heat reservoir while the small macro-system can be described in terms of thermodynamical quantities.

2. Shallow excited state.
   Only when ionized, does the acceptor have an sp^3 hybridization with a free hole in an excited state (this configuration is denoted $sp^3_- + h^+$). In this condition the acceptor atom has four covalent bonds with the same number of silicon atoms,

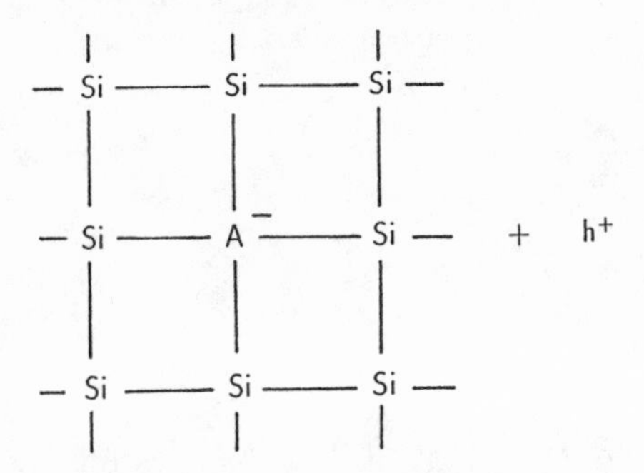

   As a consequence of the difference between their respective tetrahedral radii, the acceptor-silicon covalent bond has a different length to the silicon-silicon covalent bond. This difference will produce a strain field in the crystal, described, for instance, by the elastic continuum theory as described by *Baldereschi* and *Hopfield* [5.40].

The existence of large entropy effects for substitutional impurities in silicon and germanium due to the difference of tetrahedral radii was inferred by solid solubility data [5.41], in agreement with the analysis of section 7.3. This model will be referred to as deep dopant description (DDD) of acceptors.

Because of assumptions 1 and 2, the transition from the sp^2 to the (sp^3_- + h^+) configuration implies an elastic energy variation. Thus the chemical shift may be ascribed to the sum of two terms:

$$\Delta E = \Delta E_{\mathrm{el}} + \delta \quad ,$$

ΔE_{el} being the elastic energy change in the transition between the two configurations and δ being the electronic energy difference. In turn, ΔE_{el} can be expressed by the relationship:

$$\Delta E_{\mathrm{el}} = \frac{1}{2} k [\kappa (r - r_{\mathrm{Si}})]^2$$

where k is the elastic constant of the (sp^3_- + h^+) state and κ takes into account the fact that only a part of the difference $r - r_{Si}$ strains the lattice, while the rest is absorbed by the bond. *Baldereschi* and *Hopfield* [5.40] have calculated κ for some isovalent impurities in silicon (Ge, Sn, Pb); in this case κ may be assumed constant, $\kappa \simeq 0.4$, and we shall accept this value for all group III impurities too. The elastic constant can be evaluated from the silicon Debye temperature (T_D = 640 K) and this allows ΔE_{el} to be calculated [5.38]. The comparison of ΔE_{el} with ΔE is given in Table 5.8 and suggests an almost negligible electronic contribution δ to the chemical shift.

Table 5.8: Comparison of the elastic energy with chemical shift

Element	ΔE[eV]	ΔE_{el}[eV]
Al	0.012	0.013
Ga	0.016	0.013
In	0.099	0.119
Tl	0.189	0.148

The DDD explains the evidence in items 1 - 5 of Sect. 5.3.1 as follows:
1. Though the matter is still controversial [5.42], boron inactivation by hydrogen seems to lead to a final state where boron has an sp^2 hybridization and hydrogen saturates a silicon dangling bond:

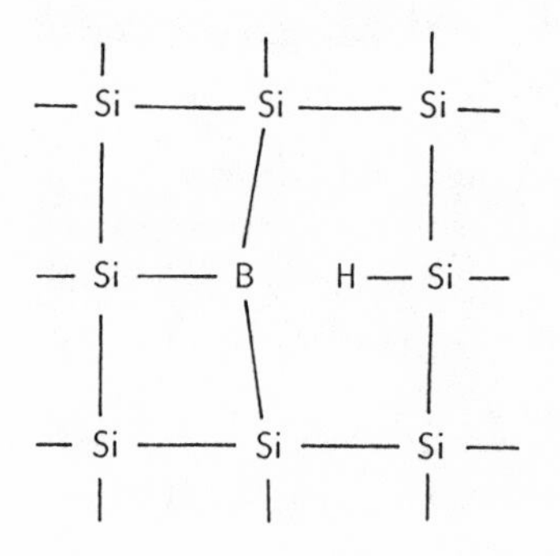

This model arose from the experiments by *Pankove* et al. [5.43], who showed that the hydrogen distribution is just the same as inactive boron and that the system exhibits the Si-H infrared absorption spectrum, and of *Stavola* et al. [5.44] who showed that hydrogen is localized close to the acceptor because at low temperature the system exhibits the A-H absorption spectrum.

In the DDD, the silicon dangling bond reacts readily with atomic hydrogen and once the dangling bond has been destroyed by reaction with hydrogen, the acceptor can no longer be ionized. This mechanism is similar to the one considered for inactivation of interface traps at the Si-SiO_2 interface after reaction with hydrogen (see Sect. 8.2):

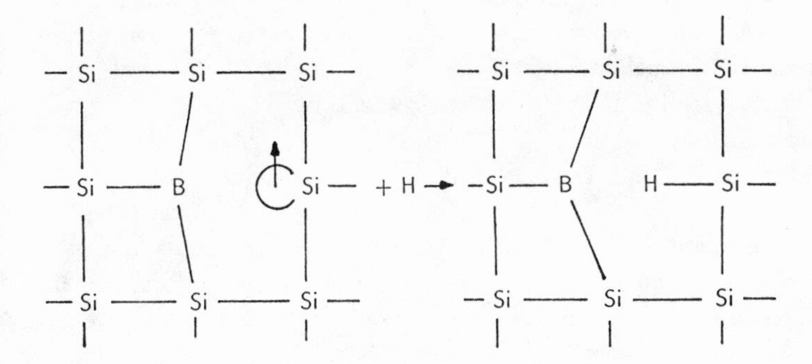

It must be noted that the STSD suggests that the acceptor is presumably passivated when it is in the ionized state (possibly through the formation of an electrostatically stabilized A^-H^+ pair), while the DDD assumes that passivation can take place only when the acceptor is in the neutral state. Strong evidence for the latter mechanism was presented by *Pankove* et al. [5.45] and by *Johnson* [5.46].

2. Indium is thermally activated with an energy of about 18 meV. The existence of this supershallow level for indium is spontaneously explained by the DDD by invoking a central role of the small macro-system [5.39].

3. In view of its low thermal ionization energy, at room temperature indium is almost completely ionized even at high concentration (up to 10^{18} cm^{-3}).

4. The piece of evidence for the Ge:Ga system suggests that one might extend the DDD of acceptors to germanium too. In fact, the behaviour of gallium in germanium is similar to that of indium in silicon.

Further results, which seem to follow from the analysis of the Si:In system (i.e. that the ionization of indium does not satisfy assumption E2, that the lattice collapses at indium concentrations higher than $10^{17}cm^{-3}$, and the consequent increase of effective mass, etc. [5.39]) still demand a confirmation.

5.4 Generation–Recombination Phenomena

In sects. 5.2 and 5.3 we have considered only equilibrium or near-equilibrium situations. A description of situations far from equilibrium, as well as of the gradual approach to equilibrium, has not been given there.

For equilibrium to be established, carrier generation (recombination) phenomena must occur to make up for carrier deficiency (excess). The theory of generation-recombination phenomena is intrinsically associated with the theory of the inhomogeneously doped semiconductor (which is able to describe complex structures such as the p-n junction and the metal-oxide-semiconductor capacitor) and hence is beyond the scope of this book. However, we recall that three mechanisms have been advocated to account for generation-recombination phenomena:

1. Shockely - Read - Hall (SRH) generation-recombination;
2. Auger recombination;
3. pure generation at donor-acceptor twins.

The SRH mechanism [5.47,48] involves an e - h pair, takes place in two consecutive steps, and requires the presence of a suitable trap T. It can be summarized by means of the consecutive reactions

$$T + e^- \rightleftharpoons T^-$$

$$T^- + h^+ \rightleftharpoons T$$

or

$$T + h^+ \rightleftharpoons T^+$$

$$T^+ + e^- \rightleftharpoons T$$

which together read

$$T + e^- + h^+ \rightleftharpoons T \quad .$$

The forward reaction is the recombination of the e - h pair at the trap, while the reverse reaction is its generation. In the overall process the energy associated with the pair is released to, or acquired by, the crystal phonon gas.

It is straightforward to observe that the efficiency of the generation process is maximum when the energy level is close to midgap. In general this occurs if the trap has an amphoteric character, such as happens for atoms with a dense electronic cloud (e.g., transition metals with filled d orbitals). SRH centres are responsible, for instance, for deviations from the Shockley description of the p-n junction [5.49], and their effects can be taken into account by more sophisticated theories [5.50].

The recombination process at deep traps is a very complex phenomenon for which the consecutive capture of individual carriers is probably an over-idealization. In particular, it seems that an important step of recombination is the formation of an exciton and hence its localization around a defect [5.51]. This implies that in (5.11) the e - h pair is bound and localized around T.

Auger recombination requires three carriers, for instance two electrons and one hole, and consists of a radiationless e - h recombination, where the energy gained in this process is acquired by the third electron. Since the probability that the minority carrier (hole) recombines increases as n^2, this process is apparent only in heavily doped silicon. The theory of Auger recombination is described for example in [5.52]; the lifetime associated with this process is given in [5.53].

When the concentration of SRH deep traps is reduced to negligible values, another mechanism becomes apparent — pure generation without recombination. This mechanism is responsible for a seemingly ohmic contribution to the reverse current of p-n junctions [5.54-56], and is field assisted and thermally activated. The activation energy of this process (of about 0.7 eV [5.55]) and the dependence on the electric field suggest that this process takes place via the emission of an e - h pair from a donor-acceptor twin (DAT). Provided that the donor-acceptor distance is in a well-defined range, the DAT behaves as a pure generation centre [5.57]. The defect forming the DAT seems to involve self-interstitials [5.58].

6. Defect-Impurity Interactions

Equilibrium defects and impurities, especially when present at high concentrations, can interact with one another. Their interactions may in turn influence remarkably the behaviour of both defects and impurities.

In principle we must consider three kinds of interactions:
- defect-defect interactions,
- impurity-impurity interactions, and
- defect-impurity interactions.

In this section 'defect' will be the short form of 'equilibrium defect'.

The defect-defect interaction has two aspects:
1) interactions between point-like defects, and
2) interactions between surface and point-like defects.

The interaction between point-like defects is responsible for the formation of: di-interstitials (i-i), di-vacancies (v-v), vacancy-interstitial pairs (v-i), larger clusters, ESFs and ISFs. The stability of the v-i pair has been discussed in Sect. 2.3. For reasons of stress, the i-i pair is supposed to be stable while the v-v pair seems unstable [6.1].

The problem of the interaction between the surface and point-like defects is very complex and often overlooked. It is usually assumed that heat treatment in an inert atmosphere always allows surface reconstruction so that at silicon surface one can tentatively impose the equilibrium concentration of vacancies and self-interstitials (see [6.2] and refs. therein quoted). We do not share this opinion and we believe that the above condition holds true only for unoxidized surfaces. With the exception of particular cases (e.g., surfaces obtained by cleavage in vacuum and kept in the same environment), it is also very questionable whether free (hence reconstructable) surfaces actually exist.

The impurity-impurity interactions may concern atoms of the *same chemical species* (e.g., clusters or precipitates) or atoms of *different chemical species* (e.g., the X centres considered later). Precipitation phenomena are usually relevant in the high density limit only, and will be considered in the next chapter; here we quote only that substitutional impurities of groups III, IV and V have solid solubilities scarcely dependent on temperature, suggesting that this quantity is limited by entropic factors. The X centres will be specifically considered in Sect. 6.3 because they can throw light on the group III acceptor puzzle. This chapter is mainly devoted to the study of defect-impurity interactions.

6.1 Defect Influence on Impurities

Information about the influence of defects on impurities usually comes from the interpretation of diffusivity experiments.

Because of the open structure of the silicon crystal, *interstitial atoms* not covalently-bonded to silicon (e.g., metals) diffuse from one site to another with relative ease, with diffusion coefficients in the range $10^{-7} - 10^{-5}$ cm^2/s at 1000 °C. If covalent bonds are involved (e.g., interstitial oxygen) the diffusion coefficient is significantly lower, of the order of 10^{-12} cm^2/s at 1000 °C.

Substitutional atoms, namely group III acceptors, group V donors, carbon and germanium are, on the contrary, characterized by much lower diffusivities, less than 10^{-14} cm^2/s at the same temperature.

About the diffusivity of *vacancies* and *self-interstitials*, many contradictory statements have been made in approximately the same period:

> "at room temperature vacancies and interstitials are mobile in silicon" [6.3];
>
> "the diffusivity of the vacancies is much higher than that of self-interstitials" [6.4];
>
> "once formed, vacancies are fairly immobile" [6.1].

In addition, the most recent experimental results and theoretical calculations give opposite results: for instance, we have already quoted the experimental results of *Taniguchi* et al. [6.5] giving a low formation energy (0.7 eV) and high migration activation energy (4.0 eV), and the extended theoretical calculations by *Car* et al. [6.6] giving a high formation energy ($5 - 8$ eV) and negligible migration barriers ($0 - 0.5$ eV). In spite of this uncertain situation, it is usually assumed that in the temperature range $800 - 1200$ °C the diffusivity of point-like defects is higher than that of substitutional impurities; for instance, the diffusivity of defects at 1000 °C is by orders of magnitude higher than 10^{-14} cm^2/s.

Various diffusion mechanisms for substitutional impurities have been proposed; the ones most frequently considered are:

1. *direct interchange* with neighbouring silicon atoms;
2. *cooperative interchange*, in which several cooperative moves occur simultaneously;
3. movement into an adjacent vacancy (*vacancy mechanism*);

4. *interstitialcy*, in which the atom occupies an interstitial site and hence moves fast until finds a vacant site.
While modes 1 and 2 are usually assumed to occur only at very high temperature because of the high energy involved in the process, modes 3 and 4 are usually assumed to occur for diffusion in the temperature range 800 – 1200 °C.

Vacancy Mechanism The vacancy mechanism is based on the following hypotheses [6.7]:
1. the impurity (El) reacts with the vacancy to form an unstable pair;
2. impurity diffusion occurs by interchange with an adjacent vacancy;
3. the impurity diffuses only when paired with a vacancy;
4. the concentration of El-v pairs is proportional to $N_{\mathrm{El}}N_{\mathrm{v}}$;
5. the vacancy concentration in each ionization state is the equilibrium one and is determined by the Fermi energy;
6. El-v pairs have different diffusivities for different vacancy ionization states.

Up to hypothesis 4, the diffusion process remains linear with the impurity concentration N_{El}; when the impurity is a dopant, however, its concentration influences the Fermi energy so that non-linearities are introduced by assumption 5. The major phenomenological basis for such an assumption is that at a given diffusion temperature T non-linearities are usually observed when the dopant concentration N_{El} exceeds the intrinsic carrier concentration $n_{\mathrm{i}}(T)$. The vacancy mechanism allows a mathematical modelling and is currently implemented in process simulation programs [6.8,9].

Interstitialcy The interstitialcy mechanism is based on the hypothesis that a substitutional atom, once injected into an interstitial position, has a high diffusivity. The conjecture of the interstitialcy mechanism stems from diffusion experiments in oxidizing environments. It has been positively demonstrated that diffusion during oxidation at temperature below 1200 °C of some dopants (e.g., phosphorus and boron) occurs with higher diffusivity than in an inert atmosphere [6.10–14]. This phenomenon is known as oxidation-enhanced diffusion. Since, in the same temperature range, oxidation injects self-interstitials (see next section), the obvious supposition is that during oxidation the equilibrium

$$\mathrm{Si_i} + \mathrm{El} \rightleftharpoons \mathrm{Si} + \mathrm{El_i} \leadsto$$

is shifted toward the rhs. The symbol $\leadsto$ [chosen in analogy with $\downarrow$ ('precipitates') and $\uparrow$ ('evaporates')] means 'diffuses fast away'. Let us consider dopants which diffuse by the vacancy mechanism; they are characterized by oxidation-retarded diffusion, so that we can also infer that self-interstitials generated during oxidation partially recombine with vacancies. Other facts which suggest a strict correlation between oxidation-enhanced diffusion and interstitial excess are discussed by *Fair* [6.15].

The actual mechanisms of dopant diffusion have long been a matter of discussion and two major schools of thought have developed; one school favours the vacancy mechanism [6.7] and the other the interstitial one [6.16]. There is now a general agreement that both mechanisms are effective in dopant diffusion, the relative weight of each mechanism being dependent on the nature of the dopant and the diffusion temperature [6.17,18]; roughly speaking, light atoms diffuse mainly by the interstitial mechanism, while heavy atoms diffuse mainly by the vacancy mechanism.

6.2 Impurity Influence on Defects

The effect of impurities on defects is usually studied by considering how impurities affect SFs — a decrease of ESF length is usually interpreted in terms of vacancy injection or interstitial absorption by impurities. No general relationship exists and in the following we shall consider a few examples which shall be useful in other parts.

Oxidation Oxidation at temperatures below 1200 °C injects self-interstitials into silicon, this conclusion being based on the following experimental observations: In most experimental situations, ESFs growing during oxidation all have the same length, which depends on the oxidation process (environment, temperature and time) [6.19]. The constancy of ESF length is interpreted by assuming that ESF nucleation sites are at the surface, and the presence or absence of ESFs is ascribed to the presence or absence of ESF nuclei. In this case one speaks of an oxidation stacking fault (OSF). The OSF lies in a (111) plane, but rather than a disc it is a semi-ellipse with major axis at the surface, because self-interstitial excess is maximum just at the surface. Figure 6.1 shows the etch pattern from SEM inspection of an OSF.

Oxygen Precipitation We have already observed that oxygen precipitation puts the silicon in a compressive state. If precipitates are large enough, the compressive energy is sufficient to form self-interstitials according to (4.1), and when self-interstitials are in large excess with respect to their equilibrium concentration they can precipitate as ESFs.

In an experimental study of oxygen precipitation kinetics in high-oxygen content silicon, *Cerofolini* and *Polignano* [6.20] observed not only ESFs, but also structures lying in (111) planes, the etch pattern of which after Secco

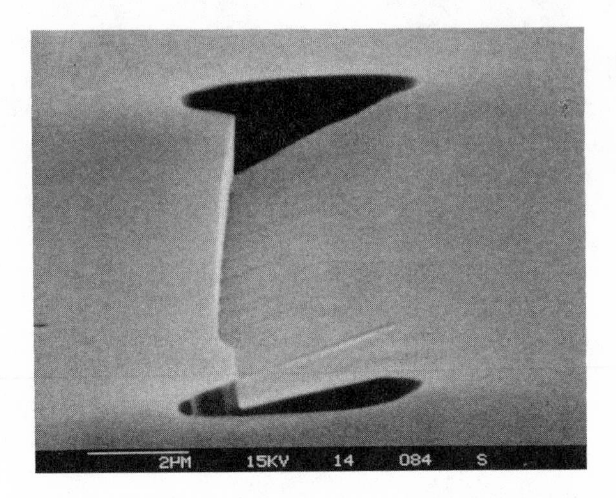

Fig. 6.1 SEM view of an OSF etch pattern after Secco etching

etching is the 'complement' of the etch pattern of the ESF (Fig. 6.2) and which was attributed to ISF. After isotropic CH_3-COOH:HNO_3:HF 5:3:1 etching to a depth of about 50 μm, it is possible to get a (100) plane crossing presumed ESFs and ISFs. The planar views of the etch pattern of these defects after Secco etching were shown in Chap. 3.

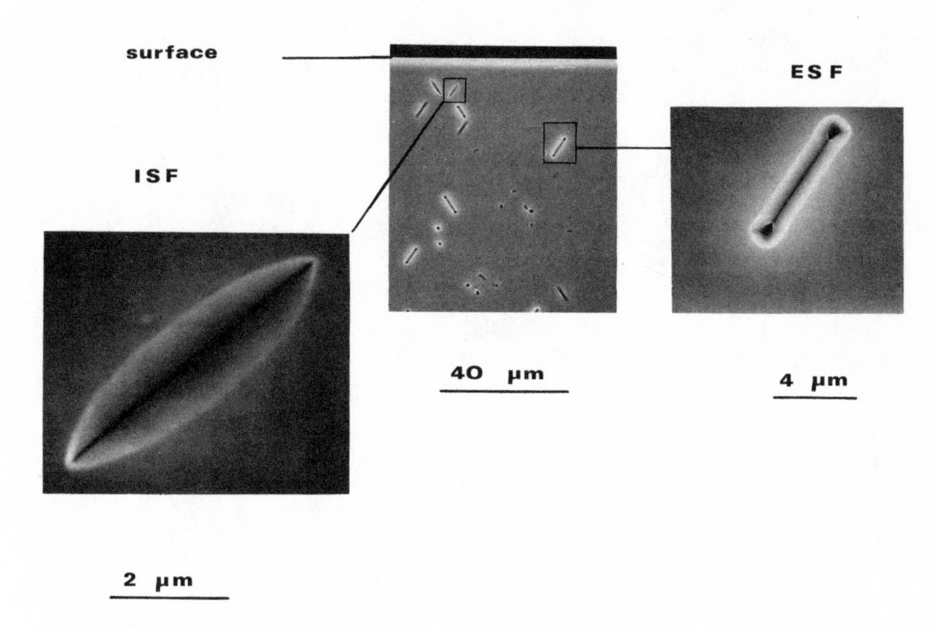

Fig. 6.2 SEM view of all the extended defects formed after an oxygen precipitation process (Secco etching after cleavage)

The complete etch pattern can be interpreted by admitting the simultaneous presence of EFSs and ISFs, in agreement with evidence for ISF formation during oxygen precipitation [6.21].

In turn, the origin of ISFs can only be explained by assuming a vacancy excess induced by precipitation. This process is possible (and mandatory) if stress is relieved in bulk silicon by injection of vacancy-interstitial pairs, rather than at the Si-$(SiO_2)_{prec}$ interface by injection of self-interstitials.

Indium Precipitation Consider an impurity for which the segregation coefficient between silicon and SiO_2 favours accumulation into silicon. If:
1) the oxidation rate is high compared to the diffusion rate, and
2) the equilibrium segregation is established at the Si-SiO_2 interface,
then oxidation piles up the dopant in silicon close to the moving interface. This 'snow-plough' effect can increase the impurity concentration to values in excess of its solid solubility.

For indium, pile-up at the Si-SiO_2 interface during oxidation was actually observed by RBS [6.22]. In that work it was also shown that:
1. samples with pre-existing precipitates showed ESFs of variable length after oxidation and Secco etching, thus showing that precipitation (which is more effective at the surface, but extends also into the bulk) produces self-interstitial excess and SF nuclei; and
2. samples without precipitates showed, after oxidation and Secco etching, a bimodal distribution of ESF length. Longer ESFs all had the same length, thus suggesting that they are OSF starting from SF nuclei at the surface; shorter ESFs, again all of the same length, indicate that at a certain stage during oxidation there is a sudden injection of self-interstitials, possibly due to indium precipitation after pile-up [6.22] or, more likely, still to a sudden lattice collapse [6.23].

6.3 Impurity-Impurity Interactions

An acceptor level involving indium, called the X centre, was observed in Hall experiments and optical absorption data by *Baron* et al. [6.24] and *Scott* [6.25], respectively. More recently, *Baron* et al. [6.26] postulated that the X centre is an In-C complex, and *Jones* et al. [6.27] have shown that each acceptor has its own X centre, i.e. each acceptor forms a relatively stable complex with carbon to give an acceptor level. Table 6.1, taken from *Jones* et al. [6.27], compares the ionization energies of all group III acceptors with those of their respective X centres.

Table 6.1: Optical ionization energies of acceptors and their X centres

Element	E_{ion}[eV]	
	pure	X centre
B	0.044	0.037
Al	0.069	0.056
Ga	0.073	0.057
In	0.156	0.113
Tl	0.246	0.180

The reason for the relative stability of the A-C complex is probably that in the ionized state the pure acceptor (except for boron) is subjected to a compressive stress, as suggested by the discussion of Sect. 5.3.2; this stress is diminished if the acceptor is bonded to an atom with tetrahedral radius smaller than r_{Si} ($r_{\text{C}} = 0.77$ Å, whereas $r_{\text{Si}} = 1.17$ Å). Through the formation of an X centre there is thus the possibility of a local stress relief.

For the X centre, too, there is the problem of explaining its chemical shift. Calculations by *Searle* et al. [6.28,29] show that the chemical shifts of X centres can be explained by a CCC-EMA. However, the DDD of acceptors is also able to explain it in a natural fashion. In this model the chemical shift of the X centre, ΔE_{AC}, is a fraction of the chemical shift of the corresponding acceptor ΔE_{A}. This fraction is probably negligible for acceptors such as Al or Ga which involve small lattice deformation (this is because the impurity can maintain the tetrahedral hybridization without lattice deformation via a modest shift towards the adjacent C atom) and this fraction must be close to $\frac{3}{4}$ for acceptors with a large tetrahedral radius (In, Tl), since in this case the sp^2 hybridization necessarily involves the compression of three bonds, while the fourth bond has the possibility of relaxing.

Table 6.2 compares the chemical shifts of acceptors, ΔE_{A}, with those of their X centres, ΔE_{AC}, and with the values $\Delta E_{\text{AC}}^{\text{DDD}}$ expected in the DDD [6.30].

Table 6.2: Chemical shifts of acceptors and their X centres.

Element	ΔE_{A}[eV]	ΔE_{AC}[eV]	$\Delta E_{\text{AC}}^{\text{DDD}}$[eV]
Al	0.012	-0.001	$\simeq 0$
Ga	0.016	0.000	$\simeq 0$
In	0.099	0.056	$\simeq 0.07$
Tl	0.189	0.123	$\simeq 0.14$

7. The High Density Limit

In this chapter we shall mainly deal with the high concentration behaviour of two classes of impurities: transition metals and impurities of the groups III, IV and V. The high concentration behaviour of oxygen has already been considered in Chap. 4.

7.1 Transition Metals

To our mind, there is no general model able to give the solid solubility of transition metals in silicon. In general, for T not too high (say, below 1200 K) all data of solid solubility C_s can be represented by the following expression

$$C_s = C_s^0 \exp(-\Delta H_s/k_B T) = N_0 \exp(\Delta S_s/k_B) \exp(-\Delta H_s/k_B T)$$

where N_0 is the atomic density of sites in which the considered atom can be dissolved. The absence of a model means that we cannot predict the values of the solution entropy ΔS_s or enthalpy ΔH_s.

For atoms which dissolve in substitutional positions (e.g., gold and platinum), one can reasonably assume that $N_0 = N_{Si}$, $\Delta S_s = \Delta S_v$ and $\Delta H_s = \Delta H_v + \Delta H_m$, where ΔH_v and ΔS_v are the vacancy formation enthalpy and entropy, respectively, and ΔH_m is the enthalpy difference between the occupied and free vacancy.

For heavy metals which dissolve as interstitials (e.g. iron and nickel), however, the dissolution requires a large negative entropy, $\Delta S_s \ll 0$, making the pre-exponential factor much lower than N_0 (in this case $N_0 \approx N_{Si}$).

The solid solubility of transition metals is influenced by dopants such as boron and phosphorus. For instance, the data of gold solubility in silicon [7.1] show that the insertion of N_P phosphorus atoms per unit volume is responsible for an increase of solid solubility from

$$C_s^{Au} = N_{Si} \exp(-\Delta H_s^{Au}/k_B T) \tag{7.1}$$

(with $\Delta H_s^{Au} = 1.7$ eV/atom) to

$$C_s^{Au} = N_{Si} \exp(-\Delta H_s^{Au}/k_B T)[1 + (N_P/N_{Si}) \exp(-\Delta H_P^{Au}/k_B T)] \quad (7.2)$$

(with $\Delta H_P^{Au} = -0.8$ eV/atom) where ΔH_P^{Au} is the change of solution enthalpy due to the insertion of phosphorus.

A comparison of (7.1) and (7.2) shows the existence of a segregation coefficient between pure and phosphorus-doped silicon

$$K_{Si}^{Si:P} = 1 + (N_P/N_{Si}) \exp(-\Delta H_P^{Au}/k_B T)$$

which tends to produce gold accumulation in the phosphorus-doped regions. This segregation coefficient is higher the lower the temperature [7.2]. Similar behaviour is observed in boron-doped silicon.

The probable reasons for a segregation coefficient which tends to accumulate metals into boron- or phosphorus-doped regions are the following:
1) because of their amphoteric character, transition metals may form ionic complexes with both donors and acceptors; for gold the complexes Au^+B^- and Au^-P^+ can form;
2) the dopant-metal pair is electrostatically stabilized (the negative ΔH_P^{Au} has a mainly Coulombic origin, as does ΔH_B^{Au});
3) the pair has no steric constraints because the tetrahedral radii of boron and phosphorus are lower than that of silicon.
In general we can hypothesize that the value ΔH_d^{Au} (d = element of groups III or V) derives from the combination of a negative electrostatic contribution tending to stabilize the pair, with a steric contribution (negative or zero for small-size dopants and positive for large-size dopants).

Other sites where transition metals tend to segregate are surfaces and extended defects. According to the general correlation proposed by *Burton* and *Machlin* [7.3], the segregation coefficient between surface and bulk is higher than unity if and only if in the solid/liquid equilibrium the liquid is richer in solute than the solid phase. Since this is actually the case for most metal-silicon binary systems (see Table 1.5), we deduce that transition metals tend to segregate at the surface and that the depth concentration profile will in general be U-shaped, i.e., there is an often unwanted accumulation of metal at the silicon surface.

Another preferential segregation site is offered by extended defects. For instance, *Tseng* et al. [7.4] gave evidence for an additional increase of the segregation coefficient in heavily phosphorus-doped silicon due to its increased density of dislocations. Clear evidence is obtained from the work of *Salih* et al. [7.5], showing that misfit dislocations at the undoped Si-Si:Ge interface in intentionally contaminated CZ silicon are decorated by gold and copper.

Another place where metals tend to segregate are ESFs, through the complex pathway outlined in Sect. 9.1. Figure 7.1 compares the TEM im-

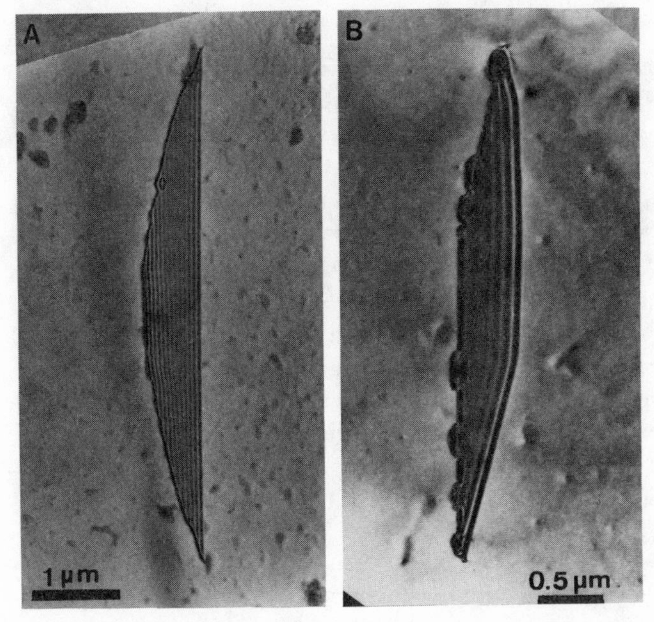

Fig. 7.1 Comparisons of TEM images of undecorated (A) and decorated (B) ESFs

ages of ESFs in FZ materials without decoration (A) and with decoration (B).

7.2 Substitutional Impurities

If substitutional impurities interact only when they are nearest neighbours, this interaction will appear with weight q $(0 < q < 1)$ when the impurity concentration N is of the order of $N_{\mathrm{Si}}/4q$. Taking, for instance, $q = 0.1$ we get $N \approx 10^{21}$ cm^{-3}.

Of course, the nearest-neighbour interaction is a very crude approximation, as the impurities may (and actually do) interact via the lattice (through electrons and phonons) and via lattice relaxation. The consideration of these factors will allow us to formulate a general model for solid solubility (Sect. 7.3.2).

Strain effects seem to play a particular role — to a first approximation, one can assume that the solid solubility is dictated by the condition that the relaxation field around each impurity may manifest freely; otherwise, any attempt to overlap different clouds of displaced atoms will produce precipitation. This argument leads to a condition on the radius Λ of the displaced cloud, $C_{\mathrm{s}} = 1/\frac{4}{3}\pi\Lambda^3$. Typical values of $C_{\mathrm{s}}(\approx 10^{20}\mathrm{cm}^{-3})$ give an estimate $\Lambda \simeq 10$ Å.

7.2.1 Clusters

Assume now one or another of the following possibilities:
1) the concentration N_1 is close to, but lower than, C_s;
2) the concentration N_2 is higher than C_s but lower than the over-saturation concentration C_{os} at which precipitation begins to occur;
3) the concentration N_3 is higher than C_{os} but the duration t of the heat treatment does not allow the formation of precipitates (that means that in the volume $\frac{4}{3}\pi(Dt)^{\frac{3}{2}}$ randomly explored by a diffusing atom there is at most, say, one other atom, $\frac{4}{3}\pi(Dt)^{\frac{3}{2}}N_3 \leq 1$).

Of course, $N_1 < N_2 < N_3$ and in none of the above situations can precipitation occur. In all cases, however, there is a non-negligible probability that impurities occupy adjacent sites. In this case we speak of clustering, and the size of the *cluster* will be the number of adjacent impurities. Dopants can conserve their own oxidation number in a cluster and also preserve stoichiometry as can be seen in the following 2D rationalizations

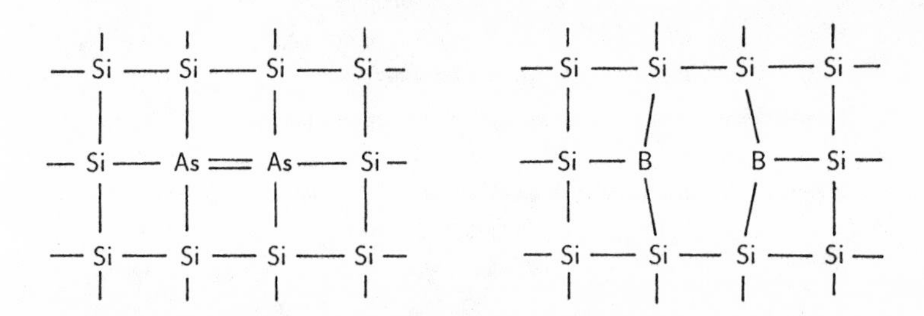

Clustering may therefore produce an electrical inactivation of dopants even when they hold substitutional positions.

Another possible cause of inactivation of substitutional impurities is their interaction with equilibrium defects. For instance, the phosphorus-vacancy pair (usually referred to as the E centre) may become inactive when the following electronic configuration is established

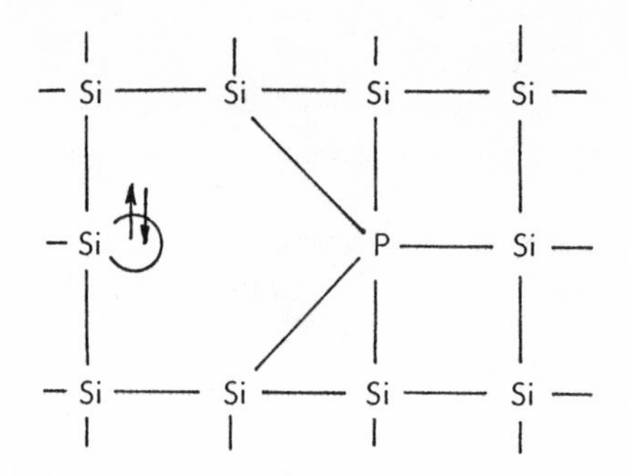

Such a configuration is the deep-centre description of the $P^{+}v^{2-}$ pair (the negatively-ionized E centre), assumed sometimes to be responsible for phosphorus inactivation [7.6,7].

7.2.2 Precipitation

When the impurity concentration exceeds the over-saturation value C_{os} and the duration of the heat treatment is long enough (condition: $\frac{4}{3}\pi(Dt)^{\frac{3}{2}}N \gg 1$) precipitation eventually occurs. The nature of precipitates depends mainly on the impurity considered. A few experimental findings are described in the following.

Carbon Carbon solid solubility is quite low, of the order of 5×10^{16} cm^{-3}, probably because of the strain field induced by its small size. Precipitation produces SiC crystals.

Germanium Germanium and silicon form a perfectly–miscible alloy. The band properties (band gap, dielectric constant, effective mass, etc.) vary regularly with composition, suggesting the possibility of modifying these properties ('bandgap engineering') by controlled growth of Si:Ge layers by molecular beam epitaxy [7.8].

Most information about dopants stems from a long-term study carried out at LAMEL. The following data are taken from EMIS data review [7.9]. The solid solubility around 1000 °C of all the dopants considered (B, P, As and Sb) is well represented by the relationship

$$C_s = C_s^0 \exp(-\Delta H_s / k_B T) \quad , \tag{7.3}$$

where ΔH_s is the solution enthalpy, and the solution entropy is contained in C_s^0. The values of parameters as well as its range of validity and the source references are reported in Table 7.1.

Table 7.1: Solid–solubility data for dopants

Dopant	C_s^0[cm^{-3}]	ΔH_s[eV]	T[°C]	Reference
B	9.25×10^{22}	0.73	900 – 1325	[7.10]
P	1.8×10^{22}	0.40	750 –1050	[7.11]
As	2.2×10^{22}	0.48		[7.12]
Sb	3.8×10^{21}	0.56	850 – 1150	[7.13,14]

A few details on the nature of precipitates are given in the following:

Boron Boron precipitates in the form of rombohedral SiB_3 [7.10].

Phosphorus Phosphorus precipitates are in the form of orthorombic SiP [7.15-17] though evidence has recently be given for coherent precipitates having the structure of cubic SiP [7.18,19].

Arsenic The precipitation of arsenic seems to occur in the form of SiAs coherent precipitates [7.20], though the matter is still controversial.

Antimony Antimony excess precipitates in the form of hexagonal Sb [7.14].

7.3 General Correlations

Two general correlations have been proposed to describe the solid solubility of various elements in silicon. In the first one the maximum solid solubility is correlated with the segregation coefficient between liquid and solid at the melting point, while in the other the solid solubility of the substitutional impurity is correlated with its tetrahedral radius.

7.3.1 Solubility and Segregation

In a paper published in 1962, *Fischler* [7.21] observed that the maximum solid solubility C_s^{max} is proportional to the segregation coefficient K at the melting point,

$$C_s^{max} = C_0 K \quad , \tag{7.4}$$

where the concentration C_0 is independent of the impurity,

$$C_0 = 5.2 \times 10^{21}\ \mathrm{cm}^{-3} \quad .$$

Observed on a log-log plot, this relationship seems to be reasonably obeyed by as many as 19 elements, including dopants, isovalent impurities, chalcogens, transition and alkali metals. Although (7.4) can be supported by thermodynamic arguments [7.22], deviations of even $2 - -3$ orders of magnitude are nonetheless observed.

The dependence of the maximum solid solubility is thus all contained in the segregation coefficient, which is strongly influenced by the impurity size. Indeed, plotting on a log-linear plot the segregation coefficient K vs the tetrahedral radius r, elements of the same group lie on the same curve given, in a first approximation, by $-\log(K/K_g) \propto r$, where K_g depends on the group while the constant of proportionality is almost independent of it. This fact was first observed by *Trumbore* [7.23]. Finschler's and Trumbore's correlations, taken together, give indications about the maximum solid solubility of an impurity with known size. In addition, since for elements of groups III, IV and V with size not too different from that of silicon, the solid solubility depends weakly on temperature, they also give an estimate of C_s in a wide temperature range. Historically, this fact has played an important role in establishing the order of magnitude of unknown solid solubilities.

7.3.2 Strain Entropy

For most physical systems the nuclear energy does not depend on the electronic energy. To a first approximation the latter is independent of the vibrational energy which, in turn, is independent of the rotational energy. Usually, these are also independent of the translational energy. When these conditions are satisfied, the free energy ΔF of the system can be decomposed as the sum of the nuclear, electronic, vibrational, rotational and translational contributions:

$$\Delta F = \Delta F_{\mathrm{nucl}} + \Delta F_{\mathrm{el}} + \Delta F_{\mathrm{vib}} + \Delta F_{\mathrm{rot}} + \Delta F_{\mathrm{trans}} \quad .$$

In solid state physics, changes of state of the system involve only changes of electronic and vibrational contributions, so that it suffices to consider only the sum

$$\Delta F = \Delta F_{\mathrm{el}} + \Delta F_{\mathrm{vib}} \quad . \tag{7.5}$$

The solid solubility in silicon and germanium of impurities that hold substitutional positions is only weakly dependent on temperature provided that their tetrahedral radii r are not too different from the one of the host matrix r' ($|r - r'|/r' < 0.2$). Figure 7.2 gives a few examples.

Starting from this observation and assuming that the factor which limits the solid solubility is purely entropic (i.e., changes of free energy in solid solution are due only to changes of entropy, $\Delta F \simeq -T\Delta S$), the entropy of the system can be decomposed as the sum of four terms,

$$\Delta S = \Delta S_{\mathrm{el}} + \Delta S_{\mathrm{vib}} + \Delta S_{\mathrm{pol}} + \Delta S_{\mathrm{str}} \quad , \tag{7.6}$$

where:

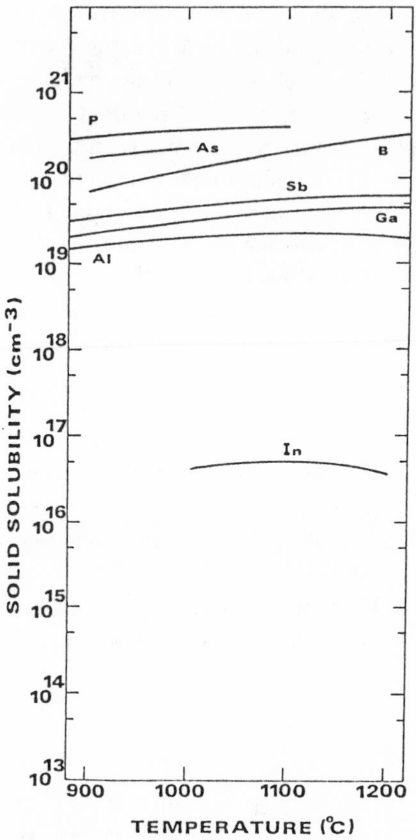

Fig. 7.2 Solid solubility of a few dopants in silicon

1) ΔS_{el} is the change of electronic entropy due to the quasi-particle (hole or electron) Fermi gas released by the substitutional impurities;
2) ΔS_{vib} is the change of vibrational entropy due to the replacement of a matrix atom (silicon or germanium) by an impurity atom; this contribution is computed by assuming an Einstein model of independent oscillators, so that the replacement of an atom does not involve any change of the vibrational entropy of the remaining atoms;
3) ΔS_{pol} is an empirical quantity, proportional to the net charge $|z|$ of the impurity ($|z| = 0$ for group IV, $|z| = 1$ for groups III and V, $|z| = 2$ for groups II and VI) and hence can be thought of as a kind of polarization contribution;
4) ΔS_{str}, the remaining contribution, was found to increase in proportion to $r - r'$ for $r > r'$ (and presumably to be negligible otherwise; this conclusion, however, being based on scanty experimental data); hence it can be thought of as a strain entropy.

Decomposition (7.6) of the solution entropy allows a rather successful description of solid solubility in silicon and germanium; in fact, about ten elements (i.e., about 10 % of the Periodic Table) obey the relationship $\Delta S_{\mathrm{str}} \propto (r - r')$ in both crystals [7.24,25]. The cost paid for this result, however, is the introduction of two new contributions, ΔS_{pol} and ΔS_{str}, not included in the general expression (7.5).

In this section we wish to study one of these strange contributions, namely ΔS_{str}, and to show that it can actually be reduced to already-known terms.

The behaviour of ΔS_{str} vs $r - r'$ for $r > r'$ clearly shows that this contribution is associated with a lattice deformation. To a first approximation, the deformed lattice can be seen as a lattice with an increased atomic density but with the same force constants (because of the negligible enthalpic contribution).

We shall assume that a zone with a different atomic density is associated with a local Debye temperature. In this view, the perturbed region must be thought of as formed by a relatively large number N_{dis} of displaced atoms in the proximity of the guest impurity. If the local Debye temperature is varied from T_{D} to $T'_{\mathrm{D}} = T_{\mathrm{D}} + \Delta T_{\mathrm{D}}$, the vibrational entropy per displaced atom is varied by an amount [7.26]

$$\begin{aligned}\frac{\Delta S_{\mathrm{vib}}}{k_{\mathrm{B}}} &= 3\left[\frac{T'_{\mathrm{D}}}{2T}\coth\frac{T'_{\mathrm{D}}}{2T} - \ln(2\ \sinh\frac{T'_{\mathrm{D}}}{2T})\right] \\ &\quad -3\left[\frac{T_{\mathrm{D}}}{2T}\coth\frac{T_{\mathrm{D}}}{2T} - \ln(2\ \sinh\frac{T_{\mathrm{D}}}{2T})\right] \\ &\simeq -3\left[\frac{T_{\mathrm{D}}}{2T}\frac{1}{\sinh(T_{\mathrm{D}}/2T)}\right]^2 \frac{\Delta T_{\mathrm{D}}}{T_{\mathrm{D}}} \quad . \end{aligned} \tag{7.7}$$

The basic idea is the following: the strain entropy per impurity atom is due to the change of vibration entropy in the displaced cloud, $\Delta S_{\mathrm{str}} = \Delta S_{\mathrm{vib}}\, N_{\mathrm{dis}}$. Together with the weak equality (7.7), this hypothesis gives

$$\frac{\Delta S_{\mathrm{str}}}{k_{\mathrm{B}}} \simeq -3\left[\frac{T_{\mathrm{D}}}{2T}\frac{1}{\sinh(T_{\mathrm{D}}/2T)}\right]^2 \frac{\Delta T_{\mathrm{D}}}{T_{\mathrm{D}}} N_{\mathrm{dis}} \quad . \tag{7.8}$$

The Debye temperature T_{D} is 640 K for silicon and 290 K for germanium, while the temperature ranges of interest are 1100 – 1400 K for silicon and 800 – 1100 K for germanium; therefore in the ranges of interest $T_{\mathrm{D}}/2T \simeq 0.3$. For $T_{\mathrm{D}}/2T < 1$ the quantity in square parentheses is very close to 1 ($x \to 0 \Rightarrow x/\sinh x \sim 1$); $-\Delta S_{\mathrm{str}}/k_{\mathrm{B}}$ is in the range 0 – 10, so that (7.8) shows that even modest changes of Debye temperature are sufficient to allow for very high strain entropy provided that N_{dis} is sufficiently high.

The change of T_D in the displaced cloud can be evaluated working in the Debye model. In fact, the cloud of displaced matrix atoms can be described, to a first approximation, as the unperturbed crystal at a somewhat different density n. The influence of n on T_D is given by

$$T_{\mathrm{D}} = \left(\frac{9}{4\pi}\right)^{\frac{1}{3}} \frac{h}{k_{\mathrm{B}}} \left(\frac{1}{c_{\mathrm{l}}^3} + \frac{1}{c_{\mathrm{t}}^3}\right)^{-\frac{1}{3}} n^{\frac{1}{3}} \quad , \tag{7.9}$$

where c_{l} and c_{t} are the longitudinal and transverse sound velocities [7.26]. These quantities can be assumed to be constant. Indeed, if the displaced cloud extends over a few atomic distances, only high frequency modes are modified and the sound velocities (related to the dispersion relationships in the zero-frequency limit) will therefore remain unchanged.
Equation (7.9) states therefore that a local increase of density, from n to $n + \Delta n$, produces thereby an increase of Debye temperature from T_D to $T_D + \Delta T_D$, where

$$\frac{\Delta T_{\mathrm{D}}}{T_{\mathrm{D}}} = \frac{1}{3}\frac{\Delta n}{n} \quad . \tag{7.10}$$

The evaluation of N_{dis} is a very difficult task because one does not know how the mismatch $r - r'$ (a kind of stress) is relaxed in the vicinity of the impurity. However, the following relationship is probably not far from reality:

$$\begin{aligned} N_{\mathrm{dis}} &\approx \frac{4}{3}\pi[\kappa(r-r')/\epsilon]^3 n \\ &= 36\pi[\kappa(r-r')]^3(n/\Delta n)^3 n \end{aligned} \tag{7.11}$$

where $\epsilon = \frac{1}{3}\Delta n/n$ is the average strain in the displaced cloud and κ is the fraction of $r - r'$ which is not absorbed by the bond itself. According to *Baldereschi* and *Hopfield's* analysis [7.27] we shall assume $\kappa = 0.4$.
Inserting (7.10) and (7.11) into (7.8) we have

$$\begin{aligned} \frac{\Delta S_{\mathrm{str}}}{k_{\mathrm{B}}} &\simeq -36\pi \left[\frac{T_{\mathrm{D}}}{2T}\frac{1}{\sinh(T_{\mathrm{D}}/2T)}\right]^2 \left(\frac{n}{\Delta n}\right)^2 [\kappa(r-r')]^3 n \\ &\simeq -36\pi \left(\frac{n}{\Delta n}\right)^2 [\kappa(r-r')]^3 n \quad . \end{aligned} \tag{7.12}$$

Since experimentally one has $\Delta S_{\mathrm{str}} \propto (r-r')$, (7.12) gives $(r-r') \propto \Delta n/n \propto \epsilon$, i.e. the average strain ϵ varies in proportion to the stress $r - r'$: the deformation remains in the elastic range.

Inserting the numerical values for silicon at 1000 °C we have $N_{\rm dis} \approx 5 \times 10^2$, $\Delta T_{\rm D}$ of the order of 1 K and the values of ϵ listed in Table 7.2

Table 7.2: Strain produced by a few substitutional impurities in silicon ($T = 1273$ K)

Element	$(r - r')[10^{-10}$ cm]	$\Delta S_{\rm str}[k_{\rm B}]$	ϵ
Al	9	4.5	2.5×10^{-3}
Ga	9	5.2	2.3×10^{-3}
In	27	9.7	8.9×10^{-3}
As	1	1.1	0.2×10^{-3}
Sb	19	4.3	7.9×10^{-3}

These results confirm that $\epsilon \propto (r - r')$, i.e. that the deformation is elastic.

Once the elastic limit is exceeded, an enthalpy term (a kind of 'deformation energy') is required for the description of solid solubility. This limit should occur at $(r - r')/r' \simeq 0.25$, and actually most systems characterized by a large enthalpy term (e.g., Bi, Au and Ag in Si, and Pb in Ge) satisfy the condition $(r - r')/r' > 0.25$.

Having reduced the contribution $\Delta S_{\rm str}$ to a possibly large perturbation to the vibrational entropy, we can hope to reduce the other strange contribution $\Delta S_{\rm pol}$ to a (large) perturbation to the electronic entropy.

8. Surfaces and Interfaces

Several important phenomena involve surfaces, some of them concerning only crystal atoms and others involving impurities too. Those concerning only crystal atoms are: surface *relaxation* (i.e., the displacement of planes close to the surface from their lattice positions) and *reconstruction* (i.e., the formation of a specific surface structure with symmetry different from that of the bulk crystal). Phenomena which involve impurities are: *segregation* at the surface; *formation of surface compounds*, this process possibly being the final result of segregation at the surface; *chemisorption* from gases or liquids, terminating at the first layer or proceeding to the formation of multilayered structures. *Oxidation* is an important example of the last process. The major technological interest is not in the silicon surface itself, but rather in the Si-SiO_2 interface, because of its very low defect density.

8.1 Amorphous SiO_2

The essential building block of all silica structure is the SiO_2 tetrahedron in which the central silicon atom has sp^3 hybridized orbitals directed towards the oxygen atoms occupying the four corners as shown in Fig. 8.1.

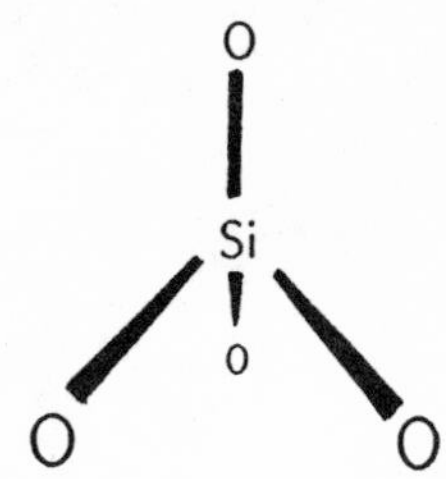

Fig. 8.1 The SiO_4 tetrahedron

In SiO_2 these tetrahedra are linked to one another by shared corners where each oxygen atom forms a bridge between two silicon atoms; in turn each silicon atom is bound to four bridging oxygen atoms. The stoichiometry is determined by the ratio of silicon and oxygen coordination numbers.

SiO_2 can be arranged in several crystalline forms, such as quartz, tridymite and coesite, but from the technological point of view the most important phase is the amorphous one. It is currently accepted that in amorphous SiO_2 tetrahedra form a continuous random network, in which the dihedral angle of the bridging oxygen is a random variable. The same structure can be seen with higher level of order, responsible for the formation of rings, consisting of a number (from 4 to 8) of tetrahedra, each of which participates in four different rings. A short account of structure, properties and defects of amorphous SiO_2 is reported in [8.1].

8.2 The Si-SiO_2 Interface

The Si-SiO_2 interface is characterized by 'proper' and 'improper' defects. Proper defects are inherently associated with the oxidation process or subsequent heat treatments; improper defects are all other defects, such as the ones due to alkali ion contamination or to radiation damage. This section will deal only with proper defects, though they are relevant only when all other defects have been reduced to a negligible amount. For instance, Na^+ contamination is responsible for a positive charge in the oxide, this charge being mobile under the action of an electric field. This mobile charge gives rise to uncontrollable shifts in the threshold voltage of metal-oxide-semiconductor (MOS) transistors which made the practical application of these devices impossible for many years. Only the use of very thorough cleaning procedures [1], oxidation in the presence of HCl (to form volatile alkali chlorides), and gettering by SiO_2:P_2O_5 (redistribution between SiO_2 and SiO_2:P_2O_5 favours Na^+ accumulation in the doped oxide) allow the mobile Na^+ charge to be reduced to a negligible level.

A silicon atom in bulk is bonded to 4 silicon atoms. Let us define the 'oxygen coordination number' of a given silicon atom as the number of oxygen atoms to which it is bonded. The region where the oxygen coordination number varies from 0 or 4 is referred to as the *interface region* between silicon and SiO_2. Representations of compounds with oxygen coordination number equal to 1, 2 or 3 are:

[1] An example of a cleaning procedure is: an extended degreasing step; a clean with a basic peroxide water solution (e.g., NH_4-OH and H_2O_2); a clean with an acidic peroxide solution (to oxidize metals); a dip with HF water solution (to remove native silicon oxides); and a rinsing with de-ionized water, with resistivity higher than 5 $M\Omega$ cm [8.1].

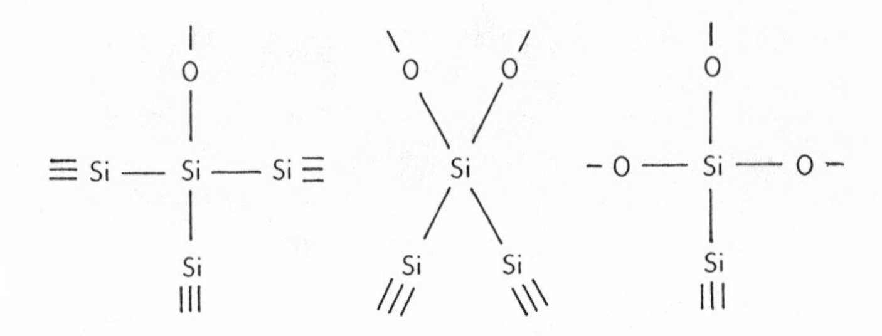

The width of the interface region has been extensively investigated; in particular:
1) *Johannessen* et al. found an upper limit of 35 Å to the interface region width [8.2];
2) *Helms* et al. lowered this limit to 20 Å [8.3]; and
3) *Helms* et al. found that this width is independent of oxide thickness x_{ox}, at least in the interval 400 Å$\leq x_{ox} \leq$ 1000 Å [8.4].

The proper defects are contained in the interface region, and are classified as *fixed charge* if they cannot exchange electrons or holes with the lattice, and as *interface traps* if their charge state can be modified by varying the Fermi energy E_F. For the amounts per unit area of fixed charge and interface traps we shall use the symbols N_f and N_{it}, respectively, as suggested by the Electrochemical Society - IEEE Committee [8.5].

Fixed Charge The fixed charge is a positive charge confined within 20 Å of the Si-SiO_2 interface. The amount of fixed charge is usually determined by measuring the shift of flat-band voltage in the high-frequency capacitance-voltage ($C - V$) characteristics of MOS capacitors [8.6]. The fixed charge depends on both oxidation-annealing and surface properties. The dependence on oxidation-annealing conditions is summarized by the *Deal* triangle [8.7] (Fig. 8.2): Starting from an oxidized surface with low N_f ($N_f \approx 10^{10}$ cm^{-2}), further oxidation at low temperature ($T \simeq 600$ °C) is responsible for an increase of fixed charge by some orders of magnitude (typically to 10^{12} cm^{-2}). Heat treatments at high temperature, at about 1200 °C both in an inert or oxidizing atmosphere, reduce N_f to values as low as 10^{10} cm^{-2}. Further annealing in an inert atmosphere at lower temperature does not yield any further increase of N_f.

This behaviour suggests that in order to obtain a thin oxide with a low fixed charge density the oxidation cycle must be carried out as follows:
1) oxidation at low T to accurately control the oxide thickness,
2) annealing at high T in an inert atmosphere to reduce N_f,
3) extraction at low T in an inert atmosphere.
The dependence of N_f on the surface structure is described by the following relation: irrespective of the oxidation-annealing conditions one finds

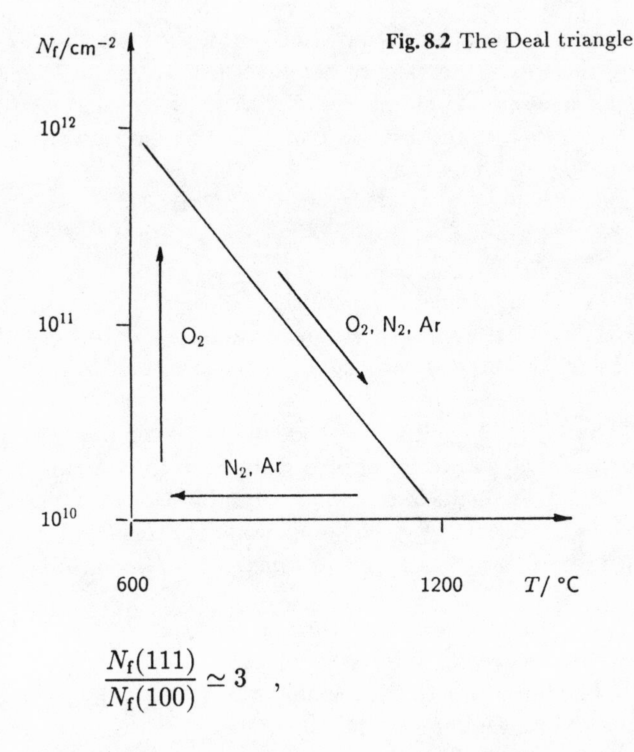

Fig. 8.2 The Deal triangle

$$\frac{N_f(111)}{N_f(100)} \simeq 3 \quad ,$$

This ratio is also roughly equal to the ratio of atomic densities on the (111) surface to the (100) surface — i.e. the higher the surface density, the higher the fixed charge.

Interface Traps Interface traps can exchange electrons or holes with the lattice. Since the interface states are characterized by a low velocity of trapping-detrapping, they can be observed by studying the quasi-static $C - V$ characteristics of MOS capacitors [8.8]. These characteristics can be used to obtain a quantity, $g_{\rm it}^T(E_{\rm F})$, related to the actual interface trap density in the gap $g_{\rm it}(E)$ through the relationship

$$g_{\rm it}^T(E_{\rm F}) = \int_{E_{\rm V}}^{E_{\rm C}} \mathcal{F}_T(E_{\rm F}, E) g_{\rm it}(E) {\rm d}E \quad , \tag{8.1}$$

where $\mathcal{F}_T(E_{\rm F}, E)$ is a function related to the Fermi - Dirac distribution [8.9]. For well prepared surfaces, the function $g_{\rm it}^T(E)$ has two maxima at about 100 meV from the conduction and valence band edges, the minimum being roughly at midgap.

The problem of solving (8.1) for $g_{\rm it}(E)$ is an improperly–posed problem, and the particular form of $g_{\rm it}^T(E_{\rm F})$ (with the maxima at the extremities of the admissible physical range) forbids even an approximate estimation of $g_{\rm it}(E)$. The experimental $g_{\rm it}^T(E)$ is however compatible with two somewhat broadened Dirac delta distributions at about 100 meV from the bottom of the conduction band and the top of the valence band, respectively.

The interface traps are generated simultaneously with the fixed charge. In fact, immediately after oxidation and irrespective of the oxidation conditions (leading to high or low N_f), one has

$$N_{it} \approx N_f \quad . \tag{8.2}$$

However, interface traps can be destroyed independently of fixed charge by heat treatments at about 450 °C in hydrogen atmosphere; heat treatments at higher temperatures (> 500 °C) regenerate the interface traps.

Structure Because of the weak inequality (8.2), one might be tempted to explain the nature of interface traps in terms of bound states in the coulombic field of the positive fixed charge. Such an explanation, however, suffers from two difficulties: it does not account for the bound state of holes and for the annealing behaviour in hydrogen which causes drastic deviations from the weak equality (8.2).

Any model for the fixed charge and interface trap must explain:
i) the localization of these defects at the interface;
ii) how they are generated simultaneously with one another;
iii) the amphoteric nature of interface traps; and
iv) why only interface traps and not fixed charges interact with hydrogen.

A model which explains the above characteristics is the following, essentially due to *Raider* and *Berman* [8.10]: Interface traps and fixed charges are formed simultaneously during oxidation by reaction of oxygen with an interface Si-Si bond

$$-O-Si-O- \; + O \longrightarrow -O-Si^{+}-O- \; + e^{-} \qquad (A)$$

The positive Si-O complex

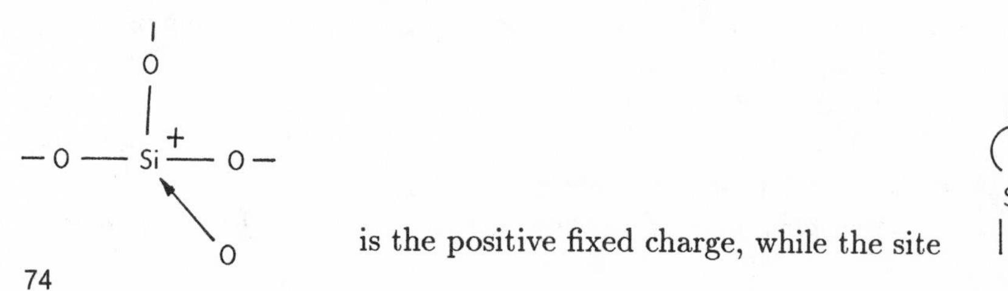

is the positive fixed charge, while the site

is the interface trap. The positive Si-O complex is assumed to be stable at low temperature (< 800 °C), while is destroyed at high temperature ($>$ 1000 °C) by the following mechanism:

```
         |                              |
         O                              O
         |+                            /
 — O —— Si —— O — + e⁻ ⟶ — O —— Si —— O —        (B)
        ↖                              \
          \                             O
           O                            |
       ↑                                Si
      (·)                               |||
       Si
       |||
```

which simultaneously destroys the interface trap. The above model explains in a natural way the features from i) to iv). In fact:

1) Both proper defects are at the interface.

2) The positive charge and interface trap are simultaneously formed (reaction A) and destroyed (reaction B).

3) The interface trap has an amphoteric character:

$$\equiv \mathrm{Si}^{\uparrow} + e^- \rightleftharpoons \mathrm{Si}^-$$

$$\equiv \mathrm{Si}^{\uparrow} \rightleftharpoons \mathrm{Si}^+ + e^-$$

The amphoteric nature of the interface traps suggests that they may act as generation-recombination centres; the location of the energy levels, close to the bands, suggests that they have a low efficiency. Actually, interface traps are responsible for a surface contribution to leakage currents of p-n junctions and MOS capacitors.

4) While the fixed charge cannot react with hydrogen, the silicon dangling bond can readily be destroyed by interaction with hydrogen at $T \simeq 450$ °C

$$\equiv \mathrm{Si}^{\uparrow} + \mathrm{H} \longrightarrow \mathrm{Si{-}H}$$

and, once saturated with hydrogen, the bond cannot further exchange electrons with the lattice. Heat treatments at $T > 500$ °C destroy the Si-H bond, as it is well known from silane chemistry.

5) The Deal triangle is explained by the assumed stability range of the positive Si-O complex.

Electronic Structure of Interface Defects Because of these assumed structures, the interface could be electron-spin resonance (ESR) active. A detailed ESR analysis of interface defects was carried out by *Caplan* et al. [8.11] and *Poindexter* et al. [8.12]. These authors oberved two intrinsic ESR signals, named P_a and P_b. The P_a signal has an isotropic character, and resembles the signal from conduction electrons. On (111) wafers the P_b signal, firstly identified by *Nishi* [8.13,14], is found to be located at the Si-SiO_2 interface; its anisotropy is very similar to that of bulk silicon defects having silicon bonded to three other silicon atoms and is in no way related to the E′ centre in SiO_2 (oxygen vacancy). In addition, both the P_b signal and N_{it} were found to be greatly reduced by steam oxidation and hydrogen annealing, while both are regenerated by subsequent annealing in a nitrogen atmosphere. These facts allow us to identify P_b with N_{it} and give a clear indication that N_{it} is therefore formed by unpaired electrons as previously sketched. The situation is less clear for (100) wafers.

The N_f centre, as hypothesized in the previous sections, is not ESR active because it is positively charged and does not contain unpaired electrons. The absence of the E′ signal associated with an oxygen vacancy in SiO_2, is therefore not in contradiction to the assumed fixed charge structure.

8.3 Oxidation Kinetics

Silicon oxidation kinetics in different environments (dry oxygen and steam) have been extensively studied. Dry oxidation is relatively slow, while steam oxidation is much faster. For fast (steam) oxidations, the kinetics (oxide thickness x_{ox} vs time t) are well described by the linear-parabolic law

$$\frac{x_{ox}^2}{k_p} + \frac{x_{ox}}{k_l} = (t + t_0) \tag{8.3}$$

due to *Deal* and *Grove* [8.15]. This law relates the SiO_2 thickness, x_{ox}, to the oxidation time t and contains three parameters: the kinetic coefficients k_p and k_l and the characteristic time t_0. The kinetic parameters define two oxidation ranges:
a) the *linear regime*, where the oxide thickness grows linearly with the oxidation time t:

$$x_{ox} \ll k_p/k_l \Rightarrow x_{ox} \simeq k_l(t + t_0), \tag{8.4}$$

and
b) the *parabolic regime*, where the oxide thickness grows parabolically with the oxidation time:

$$x_{ox} \gg k_p/k_l \Rightarrow x_{ox} \simeq [k_p(t+t_0)]^{\frac{1}{2}} \quad . \tag{8.5}$$

Kinetics of the type (8.4) are due to the fact that the rate limiting step is oxidation of interface silicon atoms, while kinetics (8.5) hold true when the rate limiting step is oxygen or hydroxyl diffusion through the oxide.

The characteristic time t_0 is related to the oxide thickness x_{ox}^0 at time $t = 0$ by

$$t_0 = x_{ox}^0/k_l + (x_{ox}^0)^2/k_p \quad ,$$

and for freshly-prepared surfaces under high vacuum $t_0 = 0$. In practice, the characteristic time t_0 is often obtained by extrapolation of actual linear kinetics (8.4) for $t \to 0$, giving $x_{ox}^0 \simeq 20 - 40$ Å, and this thickness is usually interpreted as the thickness of the 'native oxide'. This interpretation, however, is incorrect. The falsity of such an opinion is demonstrated, for instance, by neutron activation analysis and x-ray photoelectron spectroscopy showing that the oxide grown on freshly-prepared silicon after exposure to air at room temperature and pressure is lower than one monolayer [8.16]. These studies also showed that room-temperature oxidation of silicon occurs in agreement with the Elovich isotherm,

$$x_{ox} - x_{ox}^0 = r_0 \tau \ln(1 + t/\tau) \quad , \tag{8.6}$$

where τ is a characteristic time and r_0 is the growth rate at $t = 0$. The discrepancy between the common opinion about the native oxide and the above results can be overcome by assuming that (8.3) does not hold true in the first stages of oxidation.

The first attempt to modify the Deal-Grove kinetics, especially to account for the deviations observed in dry oxidation in the early stages, was contributed by *Hu* [8.17,18]: The linear oxidation regime is associated with an adsorption equilibrium of the oxidizing species described by the Heny law,

$$\vartheta \propto P \quad , \tag{8.7}$$

where ϑ is the fraction of interface sites filled by the oxidizing species, P is its partial pressure, and the constant of proportionality depends on temperature T. Equation (8.7) holds true for an energetically homogeneous distribution of adsorption sites and is seldom observed. In most practical situations the Freundlich isotherm is observed,

$$\vartheta \propto P^m \quad , \tag{8.8}$$

where m is a suitable parameter, $0 < m < 1$. Such an adsorption isotherm is related to an exponential distribution of the adsorption energy, and the parameter m is related to the width of such a distribution; a possible reason for the frequent observation of this isotherm probably resides in the fact that the exponential energy distribution is characteristic of surfaces grown in equilibrium conditions [8.19]. Interestingly enough, if the equilibrium adsorption isotherm is the Freundlich one, eq. (8.8), and one thinks of oxidation as an activated chemisorption process from an adsorbed precursor, the adsorption kinetics is given by the Elovich isotherm [8.19,20].

If the formation of an oxide and the associated diffusion-limited phenomenon are considered, one eventually arrives at the Hu kinetics, which is represented in a complicated parametric form. This oxidation isotherm behaves like the Deal - Grove isotherm (8.3) for long times, but deviates in the early stages. Hu's work will probably be the first of a series [8.21,22], since the actual heterogeneity of interface site distribution will become of increasing importance as the SiO_2 thickness is further reduced.

8.4 Surface Reconstructibility

This section is devoted to the study of self-reconstruction phenomena and their influence on equilibrium defects.

It is usually accepted that a surface under *oxidation conditions* at $T <$ 1200 °C injects self-interstitials into silicon. Self-interstitial injection must be invoked to explain ESF growth and OED.

The *nitridation process*,

$$3Si + 4NH_3 \rightarrow Si_3N_4 + 6H_2,$$

injects vacancies into silicon and in this light it can be seen as the counterpart of oxidation [8.23,24]. A detailed study of diffusion under oxidation-nitridation conditions can be found in [8.25].

For heat treatments in an *inert atmosphere*, we can distinguish two cases:

For a *free surface* (obtained, for instance, by cleavage in high vacuum and maintenance thereof) we may assume that surface reconstruction is possible (tending toward a more stable surface structure) and vacancies and self-interstitials can be injected/absorbed independently of one another, eventually reaching a concentration close to its equilibrium value in a layer of width of the order of their diffusion length. In this case, the surface can be seen as a boundary with infinite v-i generation-recombination rate and v-i profiles can be calculated by solving the Fick equation under the following boundary and initial conditions:

$$N_\mathrm{i}(0,t) = N_\mathrm{i}^\mathrm{eq} \quad , \quad N_\mathrm{v}(0,t) = N_\mathrm{v}^\mathrm{eq} \quad ,$$

and

$$N_\mathrm{i}(x,0) = \overline{N}_\mathrm{i} \quad , \quad N_\mathrm{v}(x,0) = \overline{N}_\mathrm{v} \quad ,$$

where $\overline{N}$ denotes a given concentration (e.g., the equilibrium concentration at the temperature T at which the sample was quenched). If the surface generation-recombination rate cannot be assumed infinite, equilibrium can not occur, and the boundary conditions must be modified to allow for the finite rate.

For *constrained surfaces*, we can limit ourselves to the Si-SiO_2 or Si-Si_3N_4 ones because at the moment they are of unique practical interest. In these cases the only way for vacancies and self-interstitials to independently reach equilibrium is to destroy interface bonds and eventually to reconstruct them. But this process is rather slow because of the strength of these bonds, so that it is not unreasonable to assume that in these cases vacancies and self-interstitials are generated in pairs in bulk silicon by the mechanism considered in Sect. 3.3.

Boundary Conditions First of all, we shall confine ourselves to the case of an interface in the absence of oxidation/nitridation, and consider the self-interstitial, though the following considerations can be extended to the vacancy. Let N_i be the self-interstitial concentration and consider a process at constant temperature T; the flow of self-interstitials which jump from a layer of thickness λ (of approximately an interatomic distance) to the surface is given by

$$\nu N_\mathrm{i} \lambda \exp(-\Delta E_-^* / k_\mathrm{B} T) \quad ,$$

where k_B is the Boltzmann constant, ν is the ground state vibration frequency and E_-^* is the activation energy to jump to the surface.
The flow of self-interstitials which go in solution is given by

$$n_\mathrm{s} \nu_\mathrm{s} \exp(-E_+^* / k_\mathrm{B} T) \quad ,$$

where n_s is the silicon surface density, ν_s is the vibration frequency of a surface atom, and E_+^* is the activation energy for dissolution. Of course, when an interstitial reaches the surface, or conversely, when a surface atom jumps into an interstitial site, the surface is accordingly modified, and the activation energies and ν_s are changed. In turn, the surface rearranges tending to a more stable configuration. These combined phenomena (surface change due to interstitial generation/recombination and surface reconstruction to get to an at least metastable configuration) render difficult an atomistic

description. If
1) the surface rearrangement is always faster than the interstitial generation/recombination, and
2) the local concentration is not too far from the equilibrium one N_i^{eq},
then the net generation rate can be written in the form

$$-k_i(N_i - N_i^{eq}) \quad . \tag{8.9}$$

If the surface undergoes slow variations with time, the kinetic constant k_i varies with time, and also depends on the state of the surface (free, oxidized or nitrided). Far from equilibrium (e.g., when non-thermal processes such as ion implantation are considered) equation (8.9) is not necessarily a good estimate of the generation rate. However, if we operate close to equilibrium and further assume

3) the validity of the Fick equation, and
4) the posibility of an independent (even non-thermal) flow $\mathcal{G}_i(t)$,

we get the *Hu* boundary conditions [8.26]

$$-D_i \partial N_i / \partial x = -k_i(N_i - N_i^{eq}) + \mathcal{G}_i(t) \quad .$$

9. Gettering

Silicon single crystals for semiconductor device applications are usually produced in the form of slices, with a diameter in the range 5 – 15 cm and thickness in the range 0.02 – 0.06 cm. The slice is then characterized by two major surfaces — the *front* and *back*. These surfaces have very different mechanical finishing: the front is mirror finished with extremely low roughness (peak-to-peak average distance below 30 Å), while the back is usually strongly damaged; the reasons for this finishing will become clear in the following section.

We introduce the term 'extended defect' to mean a macroscopic portion of silicon where the crystal symmetry is lost. The dimensionality, Δ, of an extended defect can be lower than 3. Extended defects may involve only silicon atoms [e.g., stacking faults ($\Delta = 2$) and dislocations ($\Delta = 1$)], impurities [e.g., precipitates ($\Delta = 3$)] or silicon-impurity complexes (e.g., the swirl defect).

Though a clear-cut, general correlation between extended defects and device electrical performances has not yet been positively demonstrated, in semiconductor device processing one usually assumes the validity of the

Aesthetic principle: beautiful = good

where 'beautiful' means 'free of extended defects' and 'good' means 'free of electrical defects'. Partial demonstrations of the validity of this assumption can be found in the observation that stacking faults (SFs) reduce the lifetime of capacitors [9.1] and increase leakage current in diodes [9.2,3].

In view of the aesthetic principle, great care is taken to avoid the growth of extended defects during processing. Though the starting material is usually provided free of dislocations and SFs, a typical device process cannot maintain this. It suffices, however, that this perfection is maintained in a limited portion of silicon, i.e. in active zones, the extension of which depends upon the kind of device considered. For instance, in MOS devices, a layer of thickness around 10 μm from the front of the slice can be considered as the active zone.

Two major techniques have been developed to remove defects — *external gettering* and *internal gettering*. These techniques are based upon very different principles, although they can be used in combination. The efficiency of a gettering technique is strongly linked with the whole process, and a high degree of empiricism is usually necessary to set up an efficient gettering technique; in the following we shall attempt a rationalization. An extended review on *Silicon material criteria for VLSI electronics* was presented by *Huff* and *Shimura* [9.4].

9.1 External Gettering

Dislocations First of all, we shall limit the discussion to dislocations; later we shall verify the influence of the process on SFs. A typical device processing procedure involves: heat treatments in inert atmospheres, oxidations, depositions of layers, implantations and definitions of geometries.

Because of the different expansion coefficients of the oxide or other layers with respect to silicon, a stress arises during a heat treatment. The level of this stress may be very high, of the order of 10^9 dyn/cm^2 and may in turn be responsible for plastic deformations. An example of plastic deformation with the formation of dislocations in Si_3N_4-masked silicon after an oxidation at high temperature is shown in Fig. 9.1.

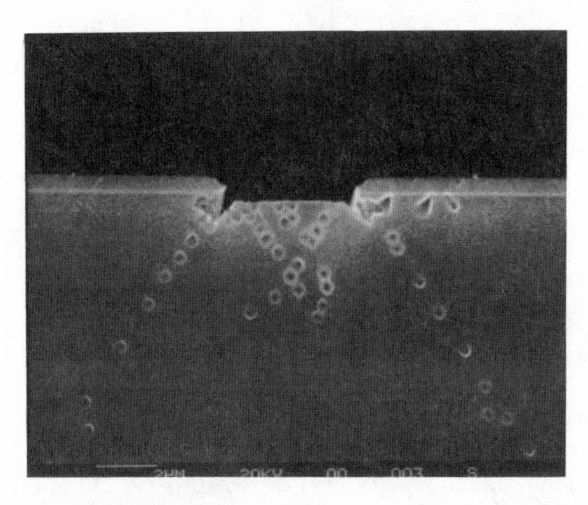

Fig. 9.1 Dislocations formed during field oxidation at the border of a Si_3N_4 mask (cleavage and Secco etching)

The plastic limits at different temperatures for FZ and CZ silicon materials are given in Fig. 1.2 [9.5]. Since the plastic limit decreases with temperature, the higher and higher stresses involved in processing larger and larger samples can only be borne without the formation of dislocation by working at low to moderate temperatures. Hence the first rule to avoid dislocations:

Rule 1: Heat treatments must be as mild as possible.

The comparison of the CZ plastic limit with that of FZ samples gives the second rule for minimizing dislocations:

Rule 2: As far as possible CZ materials are preferred to FZ.

These conditions alone, however, are not yet sufficient and other precautions must be taken.
External gettering (EG) is based upon the following

Thermodynamic conjecture: When both small and large defects are present simultaneously, heat treatments tend to enlarge large defects and to reduce (and eventually to annihilate) small defects.

This property leads to the following rule:

Rule 3: The backside must be rich in extended defects.

This condition can be satisfied starting either from a back already containing extended defects (obtained, for instance, by a local melting by laser irradiation, or by poly-silicon deposition) or from a heavily damaged back (obtained by mechanical or chemical processes) which develops extended defects immediately after the first heat treatment. Examples of heavy back damage are the stresses produced by: mechanical working (sand blasting, brushing, and so on), phosphorus predeposition (because of the difference of tetrahedral radii between phosphorus and silicon), noble gas ion implantation or Si_3N_4 deposition on the back [9.6-8]. Figure 9.2 shows two SEM pictures of two typical damaged [brushed (A) and sand-blasted (B)] backs of as–received slices.

Figure 9.3 shows the same sample backs as they are seen at SEM inspection after heavy oxidation (steam, 920°C, 6.5 h), cleavage and Secco etching. The brushed back is crowded by more, and more densely confined, dislocations than the sand-blasted one.

Stacking faults The first and third rules also help to avoid the growth of SFs. In fact, in order to organize themselves as ESFs, self-interstitials require the existence of nuclei at the surface. If the thermodynamic conjecture and rule 3 are satisfied, these nuclei tend to disappear, being gettered by extended defects at the back.

In addition, even though the gettering process of SF nuclei is not completely effective, rule 1 ensures that the length of OSFs which are formed during the growth of an oxide of given thickness is small, so that they can be dissolved with relative ease by further annealing. This conclusion is reached by considering the results of *Murarka* [9.9], who showed that for given thick-

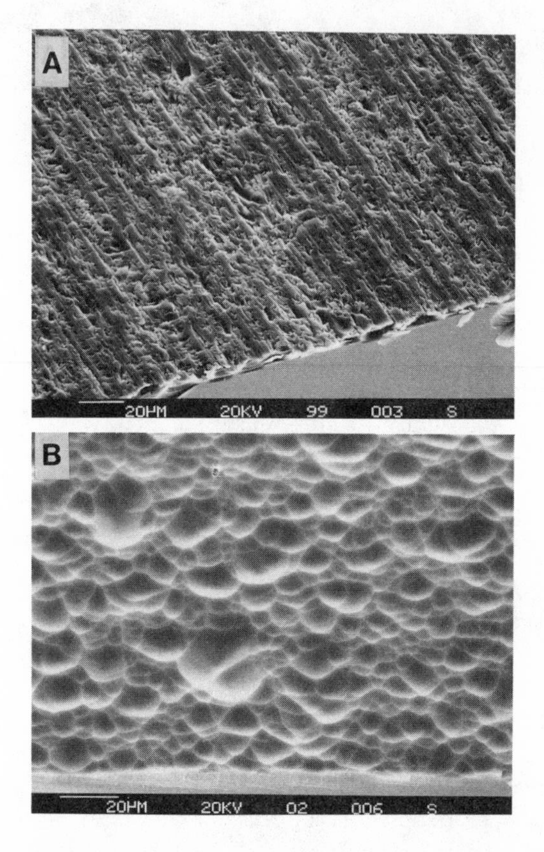

Fig. 9.2 SEM view of the backs of brushed (A) and sand blasted (B) wafers

ness of SiO_2, the length of OSFs is an increasing function of the oxidation temperature (see Table 9.1).

Table 9.1: Length of OSFs grown during a dry oxidation to produce an oxide layer of 1000 Å.

Temperature[°C]	OSF length[μm]
1050	3
1100	8
1150	15
1200	22

An additional procedure for avoiding the formation of SFs consists in carrying out oxidations, responsible for i-injection and hence likely to cause SF formation, in a HCl environment. Indeed, the following sequence seems to occur:

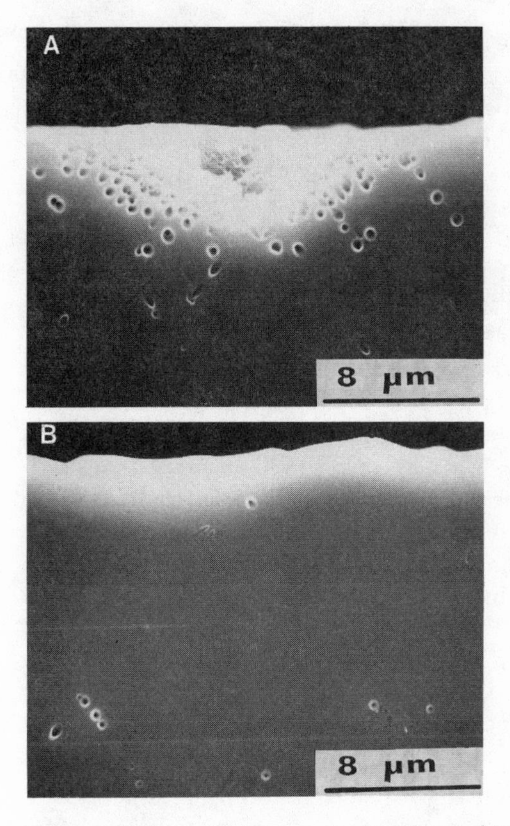

Fig. 9.3 SEM view of the same wafers (A) and (B) as in Fig. 9.2, after steam oxidation, cleavage and Secco etching. A crown of dislocations surrounds each trace of mechanical damage

Transition metal impurities tend to segregate to the surface where they may form precipitates; the precipitates, in turn, are SF nuclei and once SFs are formed they are decorated by metals. Decorated SFs are eventually responsible for electrical failures [9.10]. This vicious circle can be broken by carrying out the oxidation in HCl atmosphere, because HCl reacts with metals to form volatile chlorides and so etches away the SF nuclei.

9.2 Internal Gettering

We have already observed that in CZ materials the oxygen concentration exceeds solid solubility, even at 1200°C. If heat treatments are carried out at temperatures high enough to allow oxygen to diffuse and to overcome the over-saturation concentration, then precipitation will occur. At temperatures around 1000°C these conditions are indeed satisfied in most cases. Since important device processing steps are characterized by temperatures close to 1000°C, great care must be taken to prevent oxygen precipitation in active zones. This is achieved if the oxygen concentration therein is lowered below the over-saturation concentration. This situation can be obtained by a suitable high temperature evaporation from a region of depth comparable with the oxygen diffusion length. Such a process may also be responsible for oxygen precipitation in the bulk. This precipitation is beneficial because large oxygen precipitates in the bulk tend to dissolve small precipitates (embryos) in active regions.

In principle this high temperature process leads to the formation of a zone free of both oxygen and oxygen precipitates close to the surface (the denuded zone, DZ) and of a strongly defective zone in the bulk [9.11]. In fact, the detailed procedure to get a high quality DZ depends upon oxygen content and the thermal history of the silicon (presence of pre-existing precipitates, other defects and so on), and two major techniques are used to obtain it: the HI-LO-HI process, suitable for high oxygen concentration wafers, and the LO-HI process, suitable for low oxygen concentration wafers.

High Oxygen Content, the HI-LO-HI Process. For high oxygen content, ($8 - 10 \times 10^{17}\mathrm{cm}^{-3}$), it is mandatory to avoid oxygen precipitation in the surface layer in the early stage of processing. The first heat treatment (HI), therefore, must be done at high temperature (say, > 1050°C) to allow a fast oxygen evaporation, and so it is possible that no precipitation occurs in the bulk. To activate the precipitation process, a heat treatment (LO) at low temperature (say, around 750 °C) is carried out and is responsible for the formation of nuclei. A subsequent anneal at high temperature (HI) is responsible for the growth of precipitates [9.12]. In practice, the second HI is represented by the whole device processing; during the first HI treatment precipitation can take place by homogeneous nucleation; and during the second HI precipitation takes place by heterogeneous nucleation on pre-existing precipitates.

The effectiveness of the first HI to prevent oxygen precipitation in the surface layer is in principle increased if the process is carried out in an oxidizing environment. Indeed, oxidation injects self-interstitials into silicon so that the oxidation-induced self-interstitial excess opposes oxygen precipitation because of the action-of-mass law applied to (4.1). This inhibition is effective to a depth of the order of the self-interstitial diffusion length. Fig. 9.4 shows an example of DZ obtained after a HI-LO-HI process on medium oxygen-content silicon ($[O_i] = 6 - 8 \times 10^{17}\mathrm{cm}^{-3}$), as seen on SEM inspection after cleavage and Secco etching.

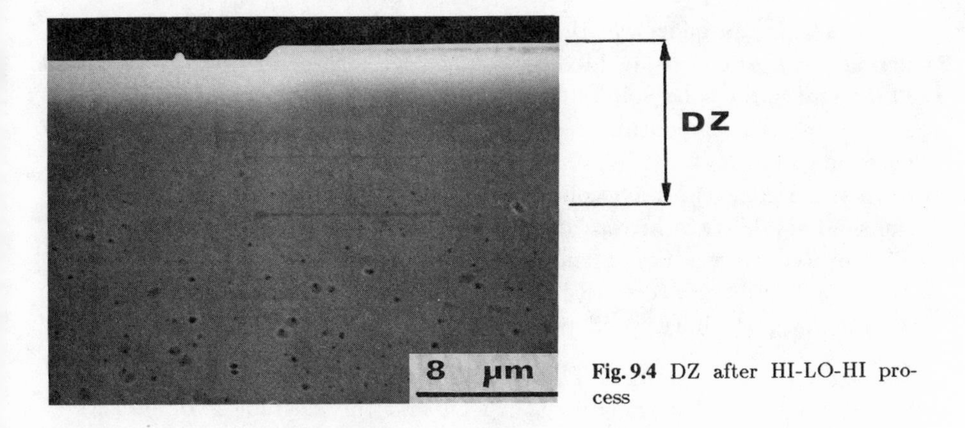

Fig. 9.4 DZ after HI-LO-HI process

Low Oxygen Content, the LO-HI Process. If the oxygen content is low ($5 - 8 \times 10^{17}$cm^{-3}), another process is suitable for DZ formation. The first step, LO, is responsible for the formation of nuclei throughout the crystal. The second step, HI, is responsible for the heterogeneous nucleation of precipitates in the bulk and for dissolution of nuclei close to the surface because the decreasing oxygen concentration makes them unstable [9.13].

9.3 Heavy-Metal Gettering

By heavy metal we mean here any metal with filled d-orbitals. Once embedded into the silicon, because of their rich electronic structure, heavy metals can exchange electrons with conduction and valence bands — i.e., they can behave as generation-recombination centres. In most cases this is an unwanted characteristic and when heavy metals are present at concentrations higher than, say, 10^{15} cm^{-3}, may be responsible for ill-functioning. Furthermore, heavy metals are fast diffusing and tend to segregate at the surface where they may precipitate because of their low solid solubility. Such metal precipitates act as ESF nuclei as described in Sect. 9.1.

Various techniques, often without a theoretical basis, have been proposed to control heavy metals; in our opinion two lines of thought can be distinguished:

Early Gettering. Metal precipitates in active zones are responsible for electrical characteristics that deviate significantly from the theoretical ones [9.14,15]; in particular, precipitates in the p-n junction are responsible for soft behaviour.

Precipitates were typical of the silicon prepared in the nineteen-fifties and sixties, because of the high metal content in silicon ingots of those

years. The major effort was then to prevent metal precipitation. Precipitation at surfaces can be avoided by:
1) increasing the solid solubility of metals, and
2) lowering the segregation coefficient between surface and bulk as close as possible to 1.
Both these goals can be accomplished if the temperature is kept high and the solid solubility is increased by heavy doping (see [9.16] and Sect. 7.1). This procedure was first identified in a classic paper by *Goetzberger* and *Shockley* [9.17] who proposed high temperature heat treatments associated with heavy phosphorus predepositions as a gettering technique.

Segregation Gettering. The steady improvement in the impurity content of commercial single crystal silicon and the cleanliness of processes have practically eliminated the problem of metal precipitation. Heavy metals, however, are often dangerous even in solid solution because, being generation-recombination centres, they behave as lifetime killers. Hence the interest in techniques able to remove heavy metals from the active zone.

One such technique, suggested for gold gettering [9.18,19], when used in semiconductor device processing [9.20] led to the realization of almost ideal p-n junctions [9.21,22] and eventually to the discovery of the pure generation mechanism [9.23].

The model of gettering by segregation annealing was presented in the communication *The getter devil and the theory of gettering* at the Los Angeles Symposium on Gettering in Semiconductors [9.24]:

The getter devil is a dwarf who, inserted into a silicon matrix, enjoys identifying and retaining heavy metal impurities walking in his neighborhood. Just as his illustrious ancestor, the Maxwell demon, he can succeed in his hobby only if the capture is a spontaneous process and if impurities can actually reach his neighborhood. 'Spontaneous' means that if capture takes place at constant temperature and volume, the process is characterized by a negative free energy difference ΔF_*° of a gettered impurity with respect to a free impurity in the silicon matrix. In this case, the spontaneous process actually occurs if the getter devil is indeed visited by the metal atoms, i.e., if the temperature is high enough.

These considerations can be formalized as follows: Let the getter devil be formed by a family of n_* suitable getter sites at an atomic concentration N_*. At any given temperature, and in equilibrium conditions there is then a preferential segregation described by the segregation coefficient

$$K = 1 + (N_*/N_{\text{Si}})\exp(-\Delta F_*^\circ/k_{\text{B}}T) \quad (9.1)$$

where T is the segregation temperature.

Relationship (9.1) shows that K increases with decreasing T and increasing N_*; these are the first two rules for gettering. The third rule is obtained by imposing that the total number of getter sites be greatly in excess of the total number of impurities, n_{imp}:

$$n_* \gg n_{\text{imp}} \quad . \tag{9.2}$$

A large value of K and the fulfillment of (9.2) are in themselves, however, insufficient for gettering. Indeed, in order to actually reach equilibrium, we must heat the crystal to render the metal impurities mobile in the matrix. From the physical point of view, we can formalize this requirement by imposing that the volume v, explored by a metal impurity during the duration t of the heat treatment, contain much more than one getter site:

$$vN_* \gg 1 \quad . \tag{9.3}$$

The theory of rate processes allows v to be estimated as

$$v = \varsigma b \nu_0 \exp(-E_{\text{m}}/k_{\text{B}}T)t \quad , \tag{9.4}$$

where ς is the cross section of a moving impurity ($\varsigma \approx 10^{-15}\text{cm}^2$), b is of the order of the silicon-silicon distance ($b \simeq 2.3$ Å), ν_0 is the vibration frequency of the metal atom in the ground state ($\nu_0 \approx 10^{13}$ s^{-1}) and E_{m} the activation energy for migration. Some values of E_{m} are the following: Ag, $E_{\text{m}} = 1.60$ eV; Au, $E_{\text{m}} = 1.12$ eV; Fe, $E_{\text{m}} = 0.87$ eV. Inserting (9.4) into (9.3) we have a condition which, for a given T [the value being fixed by the desired segregation coefficient (9.1)], specifies the required duration of the heat treatment.

The above treatment is still oversimplified, because getter sites and metal impurities are assumed to be uniformly distributed inside the slice and the probability of visiting an already visited site has not been taken into account. The latter difficulty is readily dealt with, because the number of different sites visited in long walks is a constant fraction (about 66% for simple cubic lattice) of the total number of visited sites; the former difficulty, on the other hand, may be the decisive factor in establishing gettering conditions. For instance, removing heavy metals from active zones by segregation on the back of the slice may require higher temperature or longer time than segregation in the contacts or in the scribe lines.

In the second case, a heat treatment at 800 °C for 1000 s is usually sufficient. If, however, heavy metal atoms are present in precipitates, heat treatment at moderate temperature may be insufficient. Accordingly, a preliminary treatment at high temperature, say 1100 °C, may be of help in dissolving metal precipitates [9.25].

Remembering that gettering is useful only if gettered impurities are retained far from the active regions, we can summarize the previous considerations in the following operative criteria:

1) getter sites must be formed away from active zones, for instance in the contacts or on the back;
2) their concentration must be as high as possible, and their total number must be much greater than the total number of metal impurities;
3) a higher temperature treatment must be performed, if impurities are in a precipitate form;
4) a final annealing at moderate temperature must be carried out to allow preferential segregation to getter sites.

If this procedure is directly superimposed on any standard device process before the metallization step, it represents the opposite approach to EG, where a sink of impurities is created on the back of the slice before any other process step.

Getter sites may vary greatly in their chemical and physical nature. For instance, the following structures are, or at least are said to be, getter sites: poly-silicon, phosphorus-vacancy pairs, dislocations induced by phosphorus implantation, silicon-oxygen complexes. For instance, *Tseng* et al. [9.26] showed that after a segregation anneal, the gold distribution 'copies' the phosphorus distribution, except in a dislocation-rich zone where further gold is accumulated. This example shows that both phosphorus and dislocations are effective getter sites. The gettering mechanism can vary widely from one structure to another, and no general features can be given.

Since these words, however, a lot of work has been done and we are now able to classify getter sites in relation to their effectiveness; in particular, an extended analysis involving some 10^4 junctions on a lot of substrates (p or n type, at low or high concentration) has allowed Table 9.2 to be compiled [9.27].

Table 9.2: Gettering effectiveness of dopants and extended defects

Getter site	Effectiveness		
	high	low	none
dopants	phosphorus boron		arsenic antimony
extended defects		stacking faults oxygen precipitates	

The major reason why boron and phosphorus are effective getter sites is presumably the electrostatic pairing with amphoteric impurities, as discussed in Sect. 7.1; arsenic and antimony are ineffective because their tetrahedral radii are larger than that of silicon, and this does not allow a foreign impurity to remain stably in their neighbourhoods. The gettering effectiveness of dislocations is still controversial [9.27].

A role of self-interstitials in gettering has often been hypothesized [9.28-31]; we believe that their main action is to mobilize metals through one or the other of the following mechanisms:
1. Excess self-interstitials displace substitutional metals Me (e.g., gold and platinum) into an interstitial position characterized by a very high diffusivity,

$$\mathrm{Me} + \mathrm{Si_i} \rightleftharpoons \mathrm{Si} + \mathrm{Me_i} \leadsto$$

(*kick-out mechanism* [9.32]).
2. Self-interstitials enhance the dissolution of precipitates with sub-critical radius. Because of the law of mass action, this mechanism is expected to be active for interstitial metals $\mathrm{Me_i}$ (e.g., iron, nickel) whose precipitation takes place by injection of self-interstitials:

$$\mathrm{Me_i} + (1+x)\mathrm{Si} \rightleftharpoons (\mathrm{MeSi})_{\mathrm{prec}} + x\,\mathrm{Si_i} \quad .$$

Most metals (in practice all, except iron and nickel when they are in the form of $FeSi_2$ and $NiSi_2$) participate in one or the other of the above mechanisms. Hence, irrespective of the detailed behaviour of self-interstitials, we can reasonably hypothesize that they are active in the second step of the following gettering flow-chart:

creation of getter sites
(e.g., P or B doping)

dissolution of precipitates
(e.g., self-interstitial injection
or high temperature treatment)

segregation of metal impurities to getter sites
(moderate temperature annealing)

The creation of the getter sites and the dissolution of precipitates may occur simultaneously (e.g., during a phosphorus predeposition); the segregation annealing must follow them.

9.4 Gettering and Device Processing Architecture

A device process is characterized by a number of process steps: geometry definitions, layer depositions [dielectrics (Si_3N_4 and SiO_2), metals (aluminum), semiconductors, (poly-silicon), polymers (resist)], etching, ion implantations, oxidations (dry or steam), predepositions, diffusion and anneals (inert or reducing atmospheres).

The last four processes take place at high temperature, with widely variable temperature and duration; for instance, a p- or n-well diffusion in complementary MOS involves temperatures around 1200 °C for several hours in an inert atmosphere, while the Al:Si alloy process takes place at 450 °C for about 30 min in a hydrogen atmosphere. The total number of fabrication steps in a real process usually lies between 50 and 100 (for a simplified flow chart see Sect. 10.2), about half of which involve heat treatments.

Let us restrict our analysis to MOS processes, for which most gettering techniques have been developed. A whole MOS process can be thought of as chronologically organized in 4 parts devoted to:

1. well formation at $T \simeq$ 1050–1200 °C
2. insulation and fields at $T \simeq$ 900–1000 °C
3. source and drain formation at $T \simeq$ 900 °C
4. metallization and passivation at $T \simeq$ 450 °C

Part 1 can be absent in n-channel [1] MOS processes. The heat treatments in the initial stages are much more severe than those of the latter stages, so that the dopant profile resulting after the first steps remains substantially unchanged after subsequent heat treatments. In this context, the three gettering techniques considered hold different positions: EG acts before well formation, IG starts before well formation and operates up to part 4 specified above, while gettering by segregation acts between parts 3 and 4.

[1] For reasons which will become clear in Sect. 10.1, a transistor is referred to as a p channel (n channel) if it is built in an n-type (p-type) region.

10. Device Processing

Though the transistor was invented at Bell Laboratories in 1947 (an account of *The path to the conception of the junction transistor* is given in [10.1]), the huge development of microelectronics was made possible thanks to the replacement of germanium by silicon; this replacement, rather than being due to the superior band properties of silicon compared to germanium, was fuelled by the excellent masking properties of SiO_2. These are especially evident if we compare the properties of SiO_2 with those of the oxides of the other elemental semiconductors: CO_2 is a gas while GeO_2 is water soluble. The excellent masking quality of SiO_2 [10.2] as well as the existence of a preferential etching (NH_2F_2-OH) for SiO_2 with respect to silicon and photoresist [10.3] made the ***planar technology*** possible [10.4]. Hence it was a short step to the idea of the ***integrated circuit***; this was invented by ***Noyce*** [10.5] — so decided the courts after considering the priority of several patents.

The present development, however, was driven by the MOS integrated circuit technology. Though the idea of a surface-controlled device was patented by ***Lilienfeld*** in 1930 [10.6] and ***Heil*** in 1935 [10.7], well before the development of the semiconductor theory, the MOS transistor was possible only after the availability of ultra-clean processes which allowed the high quality of the Si-SiO_2 interface to be realized. The first MOS transistor was presented by ***Khang*** and ***Atalla*** in 1960 [10.8].

The major progress from the early (metal gate, p channel, with diffused channel stoppers) MOS transistor took place through the following technological steps:

• poly-silicon deposition, which allowed source and drain electrodes to be self-aligned on the gate electrode ('silicon gate technology' [10.9-11]);
• local oxidation of silicon, which was possible after the discovery that Si_3N_4 is an excellent mask against oxidation (this technique was developed independently at Philips [10.12-14] and at SGS [10.15] and is known under several registred trademarks: LOCOS, PLANOX, isoplanar, etc.); it allows channel stoppers ('field') to be self-aligned with respect to the transistor area ('active zone');
• ion implantation, which was patented by ***Shockley*** [10.16] and brought a number of advantages: the replacement of the p-channel by the n-channel

transistor, implanted rather than diffused channel stoppers for high-voltage applications, transistor threshold-voltage control, moderate temperature processing, etc. *The evolution of silicon semiconductor technology* in the period 1952 – 1977 is reviewed in [10.17].

Each of above improvements allowed size reduction, hence better performances, higher yields and lower costs; this continuous progress has led to the current situation, in which the MOS transistor is probably the most widely produced man–made product. The total number of transistors produced up to now (end of 1987) may be estimated to be of the order of 10^{16} — and almost all of them work correctly!

10.1 The MOS Structure

In the metal-oxide-semiconductor *capacitor* (Fig. 10.1) the nature (electron or hole) and concentration of carriers can be controlled by imposing an appropriate potential at the silicon surface. This in turn is obtained by imposing a potential difference between the metal electrode (*gate*) and the bulk silicon (*body*).

According to the surface potential, the MOS capacitor can work in two typical conditions: *accumulation* (majority carriers accumulate at the surface) and *inversion* (at equilibrium, minority carriers prevail at the surface). The time required by the MOS capacitor to reach equilibrium in the inversion condition may be very long (of the order of 10^3 s at room temperature); this time is longer the lower the concentration of transition metals which behave as generation-recombination centres. The state that is obtained by a sudden application of a potential, which tends to deplete the silicon from majority carriers and which eventually evolves toward the inversion one, is referred to as *deep depletion.*

Some electron devices (e.g., the dynamic random access memory and charge-coupled device) exploit this state for their functioning; these devices require a very high holding time of the deep depletion state and hence a

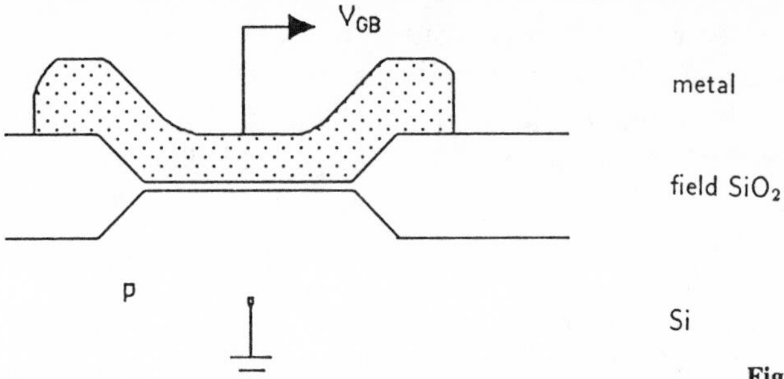

Fig. 10.1 PLANOX capacitor

very low concentration of metal contaminants. Gettering techniques have been developed mainly for such contaminants.

In other situations it is useful to operate under conditions close to equilibrium. This is achieved if the silicon region below the oxide is allowed to be almost instantaneously fed with minority carriers when they are required for the equilibrium. This is obtained by injecting the minority carriers from a suitably polarized junction. The *gated diode* (Fig. 10.2) allows this. For simplicity we have assumed a substrate of the p type, though all the considerations can be extended to the symmetric n-type configuration.

If the surface potential is negative, majority carriers accumulate and the situation is the same as for the capacitor. If however the surface potential is positive, the n^+ well is a *source* of electrons allowing the equilibrium concentration to be reached in the time allowed by carrier mobility.

If we consider now the structure of Fig. 10.3, we immediately realize that not only can we modulate the minority carrier concentration in the capacitor region by controlling the surface potential (by application of a potential V_{GS} to the capacitor electrode), but we can also drain an electron flow from one n^+ region to the other by a suitable potential difference V_{DS} between them. Because of this, the second n^+ region is referred to as the *drain* and the whole structure is the MOS *transistor*.

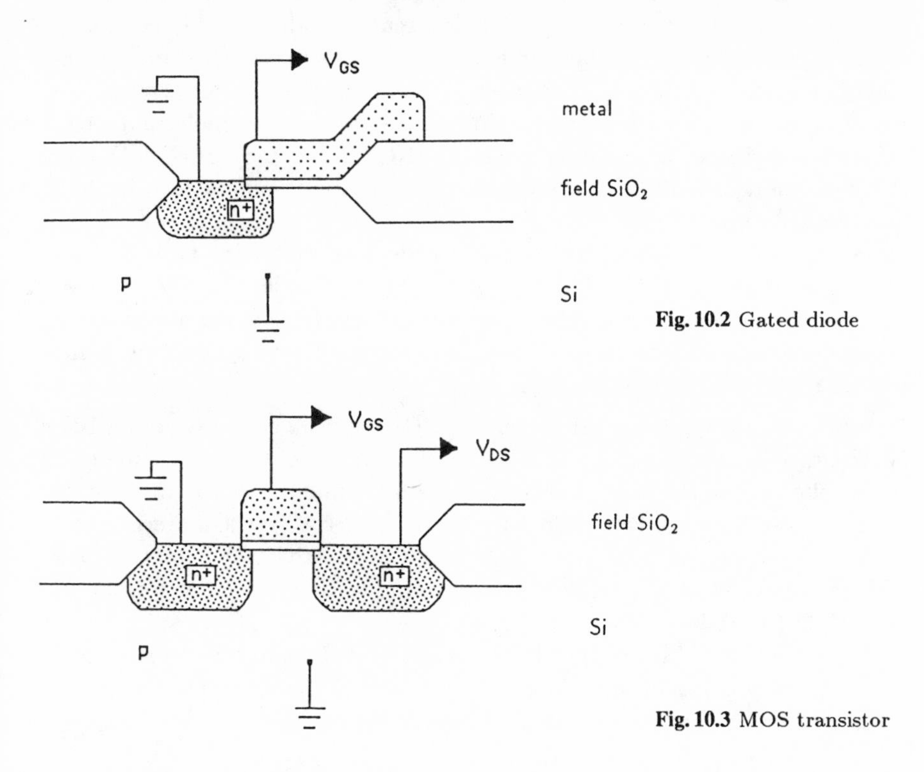

Fig. 10.2 Gated diode

Fig. 10.3 MOS transistor

Without entering into the theory of the MOS transistor [10.18], we observe that if a transistor with channel length l, channel width w and oxide thickness x_{ox} works, then a *scaled* transistor with parameters $l' = l/K$, $w' = w/K$, $x'_{ox} = x_{ox}/K$, where K is an arbitrary scaling parameter, works in the same way provided that the boundary conditions are suitably scaled [10.19]. This theorem (which is an almost straightforward consequence of the description of the electrostatic properties of silicon in terms of Poisson's equation) is only a first approximation; nonetheless, it furnishes us with a scaling criterion for producing transistors of decreasing size. Though the above criterion is expected to work best for K close to 1, it has been a useful guideline for scaling MOS transistors from $l = 4$ μm to $l = 0.25$ μm [10.20]. Scaling criteria for bipolar transistors are discussed in [10.21]. Any degree of down–scaling allows lower cost, higher yield and greater complexity — hence the continuous trend toward miniaturization.

10.2 MOS Technology

Because of the continuous progress in MOS technology, it is quite difficult to define exactly what the current technology is. Even though the microelectronics industry is rapidly becoming a mature industry, the technology is still far from the physical limits and new structures are constantly being explored in order to study their possible exploitation. As an example we mention the Si:Ge multilayered strained film (obtained by molecular beam epitaxy) for silicon band-gap engineering [10.22], and the buried SiO_2 film (by high energy, high fluence, high temperature ion implantation) for bulk device insulation [10.23].

An example of technology to produce devices with l in the range $1 - 1.5$ μm is INTEL's HMOS III technology [10.24]. Table 10.1 describes another possible process for n-channel transistors with size in the above range. We comment on a few steps of these numbered mainly by quoting which part of this book deals with them.

1. *Substrate* is p-type, CZ grown with resistivity usually in the range $10 - 50$ Ω cm. The reasons are the following: since the device performance is better the higher the minority-carrier mobility, minority carriers must be electrons and the substrate must be p-type (Sect. 5.1.3); the material is CZ to give a high plastic limit (Sect. 1.4), and the resistivity is as high as possible (compatibly with substrate inversion due to thermal donor formation, Sect. 4.2.1) to reduce area capacitance.
2. *Oxygen gettering.* For oxygen concentrations in the range $6{-}8 \times 10^{17}$ cm^{-3} a HI-LO process is suitable (Sect. 9.2). A kind of lifetime engineering can be carried out to get the desired lifetime [10.25].

Table 10.1: A typical process for producing a $1 - 1.5\ \mu m$ MOS transistor

1. Substrate: p type, CZ grown, medium oxygen content, high resistivity
2. Oxygen gettering: HI (1100 °C, 4h), LO (730 °C, 8h)
3. SiO_2 etching (diluted 1:10 HF aqueous solution)
4. Oxidation: 875°C, steam plus HCl, oxide thickness $x_{ox} = 500$ Å
5. Si_3N_4 deposition, $x_{Si_3N_4} = 1600$ Å
6. Mask (defines active zone), Si_3N_4 etching (plasma)
7. field ion implant: ^{11}B, 9×10^{12} ions/cm^2, 120 keV
8. Resist stripping (plasma)
9. Field oxidation: 920 °C steam plus HCl, $x_{ox} = 7000$ Å
10. Si_3N_4 etching (H_3PO_4), first oxide etching (HF 1:10)
11. Gate oxidation: 875°C, steam plus HCl, oxide thickness $x_{ox} = 500$ Å
12. Gate oxide annealing: 1000 °C, N_2, 15 min
13. Back oxide etching (buffered HF solution)
14. Polysilicon deposition: pyrolitic decomposition of SiH_4
15. Polysilicon doping: $POCl_3$ or P^+ ion implantation, $V/I = 5 - 6\ \Omega$
16. Mask (defines gate), poly–Si etching (plasma), SiO_2 etching (HF 1:10)
17. Resist stripping
18. Selective oxidation
19. Source and drain ion implant: ^{75}As, 6×10^{15} ions/cm^2, 150 keV
20. Radiation damage annealing: 550 °C, N_2, 2 h
21. $SiO_2 : P_2O_5$ deposition, $x_{SiO_2:P_2O_5} = 4000$ Å
22. $SiO_2 : P_2O_5$ densification, 800 °C, O_2, 35 min
23. Mask (defines contacts), $SiO_2 : P_2O_5$ etching (reactive ion etching)
24. Resist stripping
25. P predeposition , $POCl_3$, 920 °C, $V/I = 4 - 5\ \Omega$
26. Reflow anneal: 1050 °C, N_2, 5 min
27. Segregation annealing: 800 °C, N_2, 1 h
28. Al:Si evaporation, $x_{Al:Si} = 1\ \mu m$
29. Mask (defines metals), Al:Si etching (H_3PO_4)
30. Resist stripping
31. Alloying: 450 °C, H_2, 30 min
32. Back junction lapping
33. Au evaporation (on the back), $x_{Au} = 0.3\ \mu m$
34. Alloying , 400 °C, N_2, 20 min

4. 5. 6. *Active zones.* Si_3N_4 deposition takes place by reaction of SiH_4 with NH_3. The ratio of nitride thickness to oxide thickness defines the steepness of the field oxide grown in step 9 and the stress which develops during field oxidation (Sect. 9.1).
7. *Field implantation* is stopped in the active zone by the SiO_2+Si_3N_4+ resist sandwich; its role is to increase the concentration, and hence the threshold voltage, away from the active zones thus preventing parasitic MOS. It allows self-aligned channel stoppers.
9. *Field oxidation* does not occur in the active zones because these are protected by Si_3N_4 which is a mask against oxidation. The oxidation temperature follows from a compromise — it is high enough to allow viscous

flow of SiO_2 not to stress the silicon, and it is low enough to get an abrupt shape at the boundary between active zone and field.
An example showing the shape of the field oxide grown by local oxidation of silicon is shown in Fig. 10.4. The SEM picture was obtained by sectioning a cell of a 4 K dynamic random access memory. Note the complexity of the layers overlying the silicon.

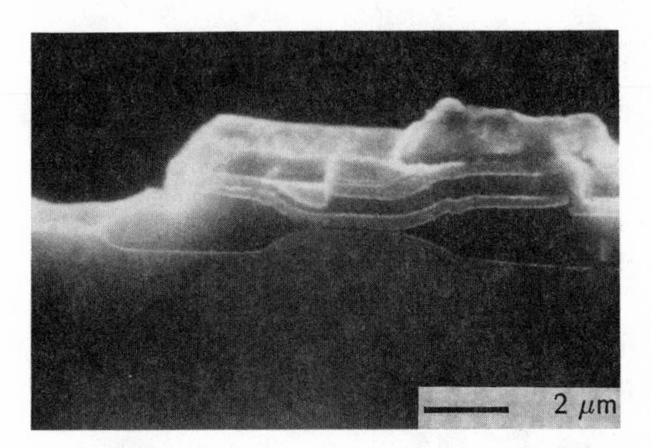

Fig. 10.4 SEM view of a cross-sectioned 4 K dynamic random-access memory cell

A few techniques to get sharper shapes of the field oxide are discussed in [10.26-28].
11. 12. *Gate oxidation and annealing* are carried out to meet the demands of low interface charge and interface states (Sect. 8.2).
18. *Selective oxidation* is carried out at low temperature to grow a thicker SiO_2 layer on the heavily doped gate region.
21-26. *SiO_2:P_2O_5 deposition and reflow* have several roles: they allow a dielectric insulation between metal and poly-silicon strips; the reflow temperature of the phosphosilicate glass ($\simeq$ 1050 °C, low compared to that of amorphous SiO_2) allows softening of the steps [10.29]; this annealing allows segregation of alkaline contaminants into the phosphosilicate (Sect. 8.2), and the temperature is high enough to dissolve metal precipitates (Sect. 9.3).
25. *Phosphorus predeposition* has different functions: it makes thicker the junction below the aluminium so preventing failures due to silicon dissolution into aluminium during the alloying (alloy spike); it furnishes the silicon with a source of getter sites (phosphorus atoms and dislocations, Sect. 9.3), and injects interstitials which act on metals as discussed in Sect. 9.3.
The structure of the contact region is shown by the TEM cross-section of Fig. 10.5; the different kinds of extended defects which are formed in the

protective SiO_2

Al:Si

SiO_2:P_2O_5
As-rich layer
field SiO_2

Si

1 μm

Fig. 10.5 TEM cross-section of a contact region in a PLANOX n^+-p junction

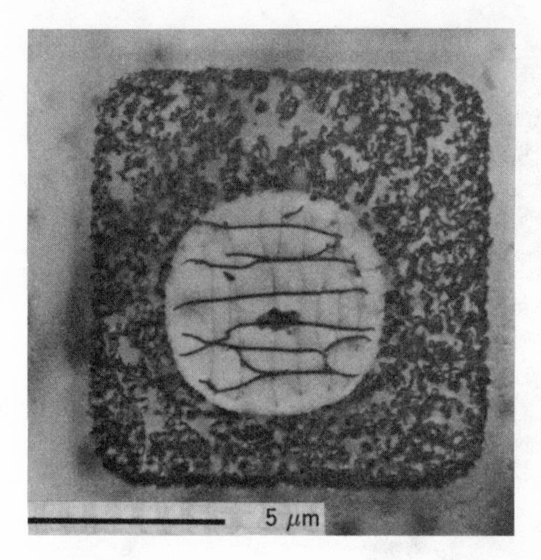

Fig. 10.6 TEM planar view of a n^+ region in a p-n junction; SFs are formed in the As-doped region and dislocations are formed in the (P + As)-doped region

As-doped regions (stacking faults) and in the (P + As)-doped regions (edge dislocations) are shown in Fig. 10.6 [10.30].

28-31. *Metal deposition and alloying.* Aluminium is doped with silicon to reduce the probability of alloy spiking [10.31]; alloying in turn supplies atomic hydrogen (from the reduction of adsorbed hydroxil groups -OH) useful for the saturation of interface traps (Sect. 8.2).

New metallization schemes are required to make up for size reduction of strip width and junction depth. The use of refractory silicides as possible replacers of poly-silicon for gates and interconnects is discussed in [10.32];

the mechanism of silicide formation on single crystal silicon is reviewed in [10.33].

A SEM image after cleavage and staining with HNO_3:HF (1%) [1] of a typical MOS transistor, which can be obtained by the technology described, is given in Fig. 10.7; details of the gate region are magnified in the lower image.

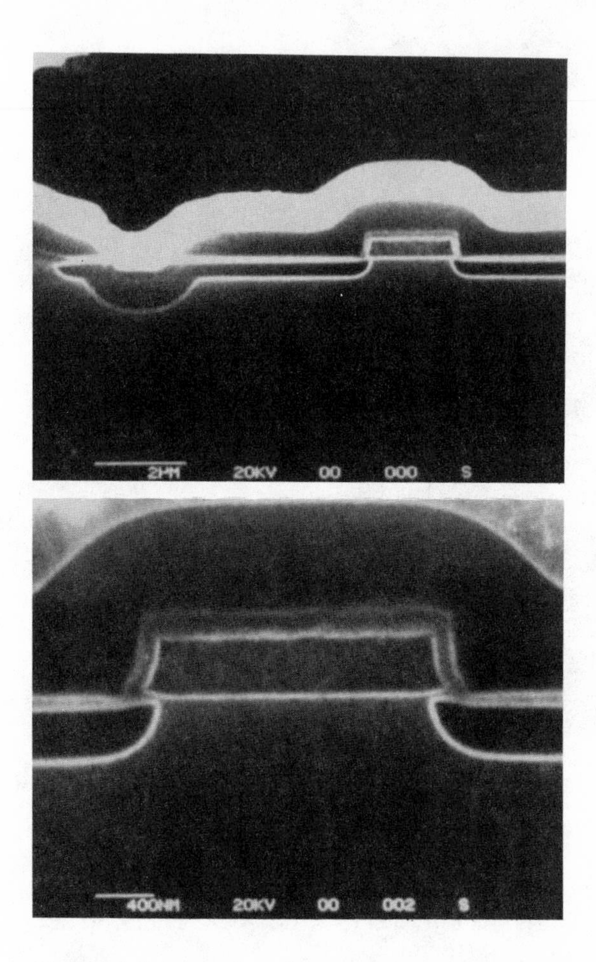

Fig. 10.7 SEM view after cleavage and staining of a transistor produced according Table 10.1. An enlarged view of the gate region is also shown

[1] HNO_3:HF (1%) is a preferential etch for n^+ silicon that allows heavily doped regions to be made visible [10.34].

10.3 A Look to the Future

The research on silicon materials is undoubtly driven by the use of silicon as the basic material for electron devices. It is likely that this activity will continue with the same intensity for a few years to come. In the following we shall discuss a few points which may well be relevant to the near future (next 10 years) of microelectronics and, hence, of silicon research.

Physical Limits It is difficult to define the physical limits met in scaling device size. They arise from a number of factors including the supply voltage (which must be higher than $k_B T/e$, where T is the operating temperature, to allow junction rectification); the oxide breakdown voltage; the channel length (below 0.1 μm the transistor description can no longer be classical, and carrier motion must be considered ballistical when size becomes of the order of the mean free path); and the intrinsic inhomogeneity of nominally equal transitors (due to statistical fluctuations of fixed charge, interface states and dopant concentrations). A few limits are discussed in [10.35-38]. Transistors of the present generation (with length in the range $0.5 - 1$ μm, oxide thickness $200 - 400$ Å, junction depth $0.1 - 0.3$ μm) are still far from most physical limits.

Complexity The trend toward complexity is summarized in the *Moore law*: the number of transistor in a chip doubles each year.

This statement, formulated as early as 1975 and applying then to the previous fifteen years [10.39], is still substantially satisfied (except for the rate of growth: the density now doubles each two years) and the trend toward greater complexity does not show saturation.

Table 10.2 gives a summary of the development of device complexity (the present technological generation is referred to as ultra large scale integration, ULSI) and a scenario up to the end of this century [10.40]. The acronyms SSI, MSI, LSI, VLSI, ULSI and HSI mean small, medium, large, very large, ultra large and 'horrendous' scale integration, respectively.

Table 10.2: Complexity of semiconductor integrated circuits

Year	Class	Bit	Transistor Length [μm]	Chip Size [mm^2]	Wafer Size [mm]
1960 – 68	SSI	2 –128	> 10	1 - 15	25
1965 – 75	MSI	64 - 4K	3 - 10	10 -25	50
1972 – 83	LSI	2K - 128K	1.5 - 4	15 -50	50 - 100
1980 – 88	VLSI	64K - 4M	0.75 - 2	25 - 75	100 - 125
1985 – 93	ULSI	2M - 128M	0.5 - 1	50 - 200	125 - 200
1990 – 99	HSI	64M - 4000M	< 0.5	100 - 400	≥ 200

Total Production Accompanying the huge increase of complexity, there is also a fast increase of the total production of silicon devices. An overall estimate of this increase can be obtained by considering the amount of single crystalline silicon produced in recent years (Fig. 10.8). The amount of silicon is given in terms of useful area, rather than of volume, because of the planar nature of integrated circuit technology.

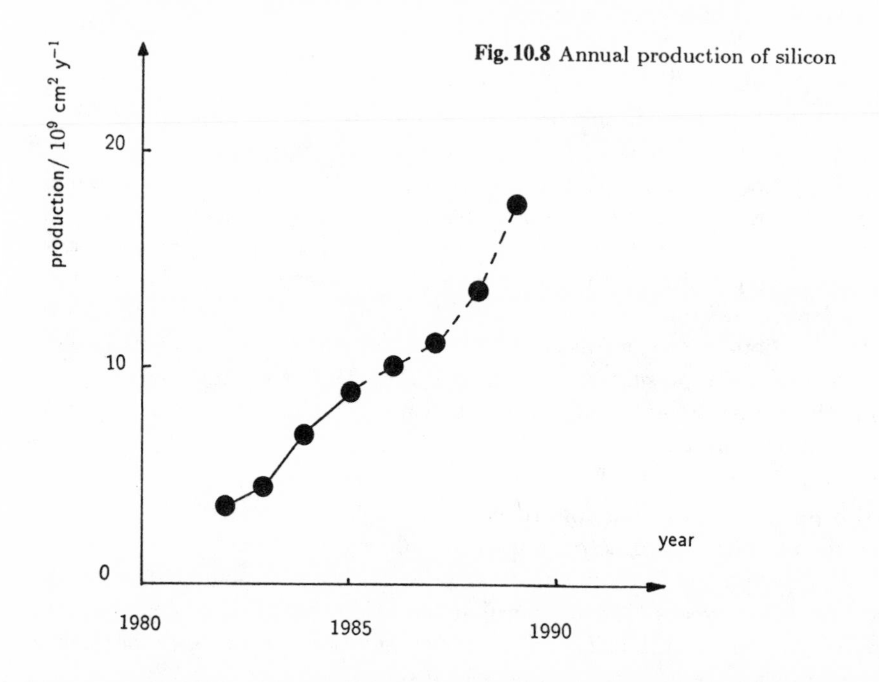

Fig. 10.8 Annual production of silicon

New Materials Despite the fact that silicon band properties are inferior to those of GaAs, certain physico-chemical properties of GaAs (non-stoichiometry of the compound at the melting point, impossibility of producing a protective layer by thermal oxidation, poor quality of all known interfaces between GaAs and deposited dielectrics) has made silicon preferable for all but a few selected applications. The techniques that can be used to make up for GaAs failures (e.g. non-stoichiometry by molecular beam epitaxy) can also be used to get new silicon-based materials, such as the Si-Si_xGe_{1-x} constrained superlattices [10.41-43] or the Si-CaF_2 heteroepitaxial interface [10.44,45].

Another way to get new interesting materials is high energy, high fluence, high temperature ion implantation to produce silicon-on-insulator (SiO_2, Si_3N_4) structures [10.23,46].

New Modes of Operation By operating at low temperature (e.g. 77 K), it is possible to reduce the supply voltage and power dissipation of integrated circuits. Operated at low temperature, silicon devices manifest new interesting properties [10.35]. The high level of activity in this new field is confirmed by the recent (1987) symposium on *Low Temperature Electronics* [10.47].

New Devices The applications of silicon are not limited to microelectronics. We quote only two new possible applications:
1. The high Debye temperature of silicon and the quality of silicon single crystals allow devices to be produced with negligible specific heat in the cryogenic regime ($T < 1$ K). These devices allow the bolometric detection of subnuclear single events [10.48]. Two kinds of applications have been hypothesized: a small device for satellite applications (astrophysical X-rays) [10.49,50] and very large devices (complete ingots) for underground detection of solar neutrinos [10.51].
2. The high radiation hardness, high purity, low reactivity and relatively low cost of silicon single crystals suggest also its possible use as shield material in plasma physics.

References

In this book only rather brief mention is made of device physics and silicon technology. For further information on device physics we suggest
A.S. Grove: *Physics and Technology of Semiconductor Devices* (Wiley, New York 1967),
S.M. Sze: *Physics of Semiconductor Devices* (Wiley, New York 1981).
Silicon technology is widely described in
R.M. Burger, R.P. Donovan: *Fundamentals of Silicon Integrated Devices Technology* (Prentice-Hall, Englewood Cliffs, NJ 1967),
S.M. Sze (ed.): *VLSI Technology* (McGraw-Hill, New York 1988).
A wide collection of data is presented in
W.R. Runyan: *Silicon Semiconductor Technology* (McGraw -Hill, New York 1965).
Several topics considered in this book are also described in our article on *Solution chemistry in silicon*, published in Intl. Rev. Phys. Chem. **7**, 123 (1988) by Francis & Taylor (1988).

Chapter 1

1.1 F.A. Cotton, G. Wilkinson: *Advanced Inorganic Chemistry* (Wiley, New York 1980) p. 375
1.2 P. Francis: *The Planets. A Decade of Discovery* (Penguin, Harmondsworth 1981)
1.3 M.P. Lepselter: Bell System Techn. J. **45**, 233 (1966)
1.4 H. Herrmann, H. Herzer, E. Sirtl: Festkörperprobleme **15**, 279 (1975)
1.5 W. Zulehner, D. Huber: Crystals **8**, 1 (1982)
1.6 J.A. Burton, R.C. Prim, W.P. Slichter: J. Chem Phys. **21**, 1987 (1953)
1.7 F. Seitz: *The Physics of Metals* (McGraw-Hill, New York 1943) Chap. 5
1.8 W.D. Sylwestrowicz: Phil. Mag. **7**, 1825 (1962)
1.9 Y. Kondo: In *Semiconductor Silicon 1977*, ed. by H.R. Huff, R.J. Kriegler Y. Takeishi (The Electrochem. Soc., Pennington NJ 1981) p. 220

1.10 K. Sumino, I. Yonenaga, M. Imai, T. Abe: J. Appl. Phys. **54**, 5016 (1983)

Chapter 2

2.1 F.A. Cotton, G. Wilkinson: *Advanced Inorganic Chemistry* (Wiley, New York 1980) p. 383
2.2 H.F. Wolf: *Silicon Semiconductor Data* (Pergamon, Oxford 1969)
2.3 M.T. Yin, M.L. Cohen: Phys. Rev. Lett. **45**, 1004 (1980)
2.4 M.T. Yin, M.L. Cohen: Phys. Rev. B **26**, 5668 (1982)
2.5 R.J. Needs, R.M. Martin: Proc. 17th Intl. Conf. Physics of Semiconductors (1985), p. 965
2.6 V.G. Eremenko, V.I. Nikitenko: Phys. Status Solidi (a) **14**, 317 (1972)
2.7 T.Y. Tan, H. Föll, S.M. Hu: Phil. Mag. A **44**, 127 (1981)
2.8 G.F. Cerofolini, L. Meda, G. Queirolo, A. Armigliato, S. Solmi, F. Nava, G. Ottaviani: J. Appl. Phys. **56**, 2981 (1984)
2.9 G.F. Cerofolini, P. Manini, L. Meda, G.U. Pignatel, G. Queirolo, A. Garulli, E. Landi, S. Solmi, F. Nava, G. Ottaviani, M. Gallorini: Thin Solid Films **129**, 111 (1985)
2.10 M. Servidori, S. Cannavò, G. Ferla, A. La Ferla, S.U. Campisano, E. Rimini: Nucl. Instrum. Meth. B **19/20**, 317 (1987)
2.11 G.F. Cerofolini, L. Meda, M.L. Polignano, G. Ottaviani, A. Armigliato, S. Solmi, H. Bender, C. Claeys: *Semiconductor Silicon 1986*, ed. by H.R. Huff, T. Abe, B. Kolbesen (The Electrochem. Soc., Pennington NJ 1986) p. 706
2.12 A. Bourret: Inst. Phys. Conf. Ser. **87**, 39 (1987)
2.13 L. Meda, G.F. Cerofolini, G. Ottaviani: Nucl. Instrum. Meth. B **19/20**, 454 (1987)
2.14 G.F. Cerofolini, L. Meda: Phys. Rev. B **36**, 5131 (1987)
2.15 P. Sigmund, J.B. Sanders: Proc. Int. Conf. on Applications of Ion Beams to Semiconductor Technology, ed. by P. Glotin (OPHRYS 1967) p. 215
2.16 J.F. Gibbons: Proc. IEEE **59**, 1063 (1972)
2.17 D.A. Thompson, A. Golanski, K.H. Hauger, D.V. Stefanovic, G. Carter, C.E. Chrestoulides: Rad. Eff. **52**, 69 (1980)
2.18 S. Prussin: 167th Meeting Electrochem. Soc. (1985) abs. no. 255
2.19 A.M. Mazzone: IEEE Trans. Computer Aided Design **CAD-4**, 110 (1985)
2.20 G.F. Cerofolini, L. Meda, G. Ottaviani: Nucl. Instrum. Meth. B **19/20**, 488 (1987)

2.21 M. Geddo, D. Maghini, A. Stella: Solid State Commun. **58**, 483 (1986).
2.22 M. Servidori, R. Angelucci, F. Cembali, P. Negrini, S. Solmi, P. Zaumseil, U. Winter: J. Appl. Phys. **61**, 1834 (1987)
2.23 L. Csepregi, E.F. Kennedy, T.J. Gallagher, J.W. Mayer, T.W. Sigmon: J. Appl. Phys. **48**, 4234 (1977)
2.24 G.K. Hubler, C.N. Waddell, W.G. Spitzer, J.E. Fredrickson, T.A. Kennedy: Mat. Res. Soc. Symp. Proc. **27**, 217 (1984)
2.25 C. Claeys, H. Bender, G.F. Cerofolini, L. Meda: 171st Meeting Electrochem. Soc. (1987) abs. no. 222
2.26 L. Csepregi, E.F. Kennedy, S.S. Lau, J.W. Mayer, T.W. Sigmon: J. Appl. Phys. **29**, 645 (1976)
2.27 N.F. Mott: Contemp. Phys. **18**, 225 (1977)
2.28 S.R. Ovshinky, D. Adler: Contemp. Phys. **19**, 83 (1978)
2.29 A. Madan, P.G. Le Comber: W.E. Spear, J. Non-Cryst. Solids **11**, 219 (1976)
2.30 E.P. Donovan, F. Spaepen, D. Turnbull, J.M. Poate, D.C. Jacobson: Appl. Phys. Lett. **42**, 698 (1983)
2.31 E.P. Donovan, F. Spaepen, D. Turnbull, J.M. Poate, D.C. Jacobson: J. Appl. Phys. **57**, 1795 (1985)
2.32 J.C.C. Fan, H. Andersen: J. Appl. Phys. **52**, 4003 (1981)
2.33 A. Ourmazd, K. Ahlborn, K. Ibeh, T. Honde: Appl. Phys. Lett. **47**, 685 (1985)
2.34 A. Ourmazd, J.C. Bean, J.C. Phillips: Phys. Rev. Lett. **55**, 1599 (1985)
2.35 J.C. Phillips, J.C. Bean, B.A. Wilson, A. Ourmazd: Nature **325**,121 (1987)

Chapter 3

3.1 G.D. Watkins: Inst. Phys. Conf. Ser. **23**, 1 (1975)
3.2 G.D. Watkins: Inst. Phys. Conf. Ser. **46**, 16 (1979)
3.3 B.J. Masters, E.F. Gorey: J. Appl. Phys. **49**, 2717 (1978)
3.4 R. Car, P.J. Kelly, A. Oshiyama, S.T. Pantelides: Phys. Rev. Lett. **52**, 1854 (1984)
3.5 J. Van Vechten: Phys. Rev. B **10**, 1482 (1974)
3.6 J. Van Vechten: Proc. 13th Intl. Conf. Defects in Semiconductors (The Metallururgical Soc., Warrenale PA 1985) p. 293
3.7 A. Seeger, W. Frank: Proc. 13th Intl. Conf. Defects in Semiconductors (The Metallururgical Soc., Warrendale PA 1985) p. 159
3.8 H.M. James, K. Lark-Horovitz: Z. Phys. Chem. (Leipzig) **198**, 107 (1951)

3.9 W. Frank: Inst. Phys. Conf. Ser. **23**, 23 (1975)
3.10 H.J. Mayer, H. Mehrer, K. Maier: Inst. Phys. Conf. Ser. **31**, 186 (1977)
3.11 K. Taniguchi, D.A. Antoniadis, Y.Matsushita, Appl. Phys. Lett. **42**, 961 (1983)
3.12 G.B. Bronner, J. Plummer: 166th Meeting Electrochem. Soc. (1984) abs. no. 483
3.13 S.T. Pantelides: In *The Physics of VLSI*, ed. by J.C. Knights (Am. Inst. Phys., New York 1984) p. 125
3.14 G.K. Wertheim: Phys. Rev. **115**, 568 (1959)
3.15 S.M. Hu: J. Vacuum Sci. Technol. **14**, 17 (1977)
3.16 U. Gösele, W. Frank, A. Seeger: Solid State Commun. **45**, 31 (1982)
3.17 C. Kittel: *Introduction to Solid State Physics* (Wiley, New York 1976) problem 17.1
3.18 J. Desseaux-Thibault, A. Bourret, J.M. Peuisson: Inst. Phys. Conf. Ser. **67**, 71 (1983)
3.19 G.F. Cerofolini, L. Meda, M.L. Polignano, G. Ottaviani, A. Armigliato, S. Solmi, H. Bender, C. Claeys: *Semiconductor Silicon 1986*, ed. by H.R. Huff, T. Abe, B. Kolbesen (The Electrochem. Soc., Pennington NJ 1986) p. 706
3.20 I.L.F. Ray, D.H.J. Cockayne: Proc. Roy. Soc. (London) A **235**, 543 (1971)
3.21 H. Föll, C.B. Carter: Phil. Mag. A **40**, 497 (1979)
3.22 M.Y. Chou, S.G. Louie, M.L. Cohen: Proc. 17th Intl. Conf. Physics of Semiconductors (1985) p. 43
3.23 F. Secco D'Aragona: J. Electrochem. Soc. **119**, 948 (1972)
3.24 W. Zulehner, D. Huber: Crystals **8**, 1 (1982)
3.25 C. Claeys, H. Bender, G. Declerck, J. Van Landuyt, R. Van Overstraeten, S. Amelinckx: In *Aggregation Phenomena of Point Defects in Silicon*, ed. by E. Sirtl, J. Gorissen (The Electrochem. Soc., Pennington NJ 1984) p. 74
3.26 G.F. Cerofolini, M.L. Polignano: *The Physics of VLSI*, ed. by J.C. Knights (Am. Inst. Phys., New York 1984) p. 225
3.27 M.L. Polignano, G.F. Cerofolini, H. Bender, C. Claeys: *ESPRIT '85*, ed. by the Commission of the European Communities (North Holland, Amsterdam 1986) Part 1, p. 233

Chapter 4

4.1 B.O. Kolbesen, A. Mülbauer: Solid State Electron. **25**, 759 (1982)
4.2 W. Zulehner, D. Huber: Crystals **8**, 1 (1982)

4.3 G.S. Oehrlein, J.W. Corbett: Mat. Res. Soc. Symp. Proc. **14**, 107 (1983)
4.4 C. Plougoven, B. Leroy, J. Arhan, A. Lecuiller J. Appl. Phys. **49**, 2711 (1978)
4.5 S.M. Hu: Appl. Phys. Lett. **48**, 115 (1986)
4.6 S.M. Hu: J. Appl. Phys. **52**, 3974 (1981)
4.7 V. Cazcarra, P. Zunino: J. Appl. Phys. **51**, 4206 (1980)
4.8 W. Kaiser, H.L. Frisch, H. Reiss: Phys. Rev. **112**, 1546 (1958)
4.9 D. Helmreich, E. Sirtl: In *Semiconductor Silicon 1977*, ed. by H.R. Huff, E. Sirtl (The Electrochem. Soc., Princeton NJ 1977), p. 626
4.10 P. Rava, H.C. Gatos, J. Lagowski: In *Semiconductor Silicon 1977*, ed. by H.R. Huff, R.J. Kriegler, Y. Takeishi (The Electrochem. Soc., Pennington NJ 1981), p. 232
4.11 G.D. Watkins, J.W. Corbett: Phys. Rev. **121**, 1001 (1961)
4.12 U. Gösele, T.Y. Tan: Appl. Phys. A 28, 79 (1982)
4.13 b. Bourret: Proc. 13th Intl. Conf. Defects in Semiconductors (The Metallururgical Soc., Warrendale PA 1985) p. 129
4.14 A. Kanamori, M. Kanamori: J. Appl. Phys. **50**, 8095 (1979)
4.15 Y. Matsushita: Proc. 17th Intl. Conf. Physics of Semiconductors (1985) p. 1525
4.16 A. Bourret: Inst. Phys. Conf. Ser. **87**, 39 (1987)
4.17 A. Bourret, J. Thibault-Desseaux, D.N. Seidman: J. Appl. Phys. **55**, 825 (1984)
4.18 F. Shimura, R.A. Craven: In *The Physics of VLSI*, ed. by J.C. Knights (Am. Inst. Phys., New York 1984) p. 205

Chapter 5

5.1 M.L. Cohen, T.K. Bergstresser: Phys. Rev. **141**, 789 (1966)
5.2 Y. Wu, L.M. Falicov: Phys. Rev. B **29**, 3671 (1984)
5.3 C. Kittel, A.H. Mitchell: Phys. Rev. **96**, 1488 (1954)
5.4 J.M. Luttinger, W. Kohn: Phys. Rev. **97**, 969 (1955)
5.5 W. Kohn: Solid State Phys. **5**, 257 (1957)
5.6 S.T. Pantelides: Rev. Mod. Phys. **50**, 797 (1978)
5.7 M. Lannoo, J. Bourgoin: *Point Defects in Semiconductors I*, Springer Ser. Solid–State Sci. **22**, (Springer, Berlin, Heidelberg 1981)
5.8 P.P. Edwards, M.J. Sienko: Phys. Rev. B **17**, 2575 (1978)
5.9 E.A. Guggenheim: *Thermodynamics* (North Holland, Amsterdam 1967) ch. 7
5.10 C.S. Fuller: Rec. Chem. Progr. **17**, 75 (1956)

5.11 P.M. Solomon: In *The Physics of VLSI*, ed. by J.C. Knights (Am. Inst. Phys., New York 1984) p. 172
5.12 L. Reggiani, Proc. 15th Intl. Conf. Physics of Semiconductors (1980) p. 317
5.13 B.R. Nag: *Theory of Electrical Transport in Semiconductors* (Pergamon, London 1972)
5.14 R.D. Larrabee, W.R. Thurber, W.M. Bullis: *Semiconductor Measurement Technology* (NBS Special Publication 400-63, Washington 1980)
5.15 P. Cappelletti, G.F. Cerofolini, G.U. Pignatel: J. Appl. Phys. **53**, 6457 (1982)
5.16 G. Masetti, M. Severi, S. Solmi: IEEE Trans. Electron Dev. **ED-30**, 764 (1983)
5.17 A.G. Milnes: *Deep Impurities in Semiconductors* (Wiley, New York 1973) ch. 1
5.18 A. Morita, H. Nara: Proc. 8th Intl. Conf. Physics of Semiconductors (1966) p. 234
5.19 C.E. Jones, D. Schafer, W. Scott, R.J. Hager: J. Appl. Phys. **52**, 5148 (1981)
5.20 T.N. Morgan: Proc. 10th Intl. Conf. Physics of Semiconductors (1970) p. 266
5.21 H.F. Wolf: *Silicon Semiconductor Data* (Pergamon, Oxford 1969)
5.22 S. Shinohara: Nuovo Cimento **22**, 18 (1961)
5.23 N.O. Lipari, M.L.W. Thewalt, W. Andreoni, A. Baldereschi: Proc. 15th Intl. Conf. Physics of Semiconductors (1980) p. 165
5.24 C.T. Sah, J.Y.C. Sun, J.J. Tzou: Appl. Phys. Lett. **43**, 204 (1983)
5.25 J.I. Pankove, D.E. Carlson, J.E. Berkeyheiser, R.O. Wance: Phys. Rev. Lett. **51**, 2224 (1983)
5.26 R. Gale, F.J. Feigl, C.W. Magee, D.R. Young: J. Appl. Phys. **54**, 6938 (1983)
5.27 C.T. Sah, J.Y.C. Sun, J.J. Tzou: J. Appl. Phys. **55**, 1525 (1984)
5.28 J.I. Pankove, R.O. Wance, J.E. Berkeyheiser: Appl. Phys. Lett. **45**, 1100 (1984)
5.29 W.L. Hansen, S.J. Pearton, E.E. Haller: Appl. Phys. Lett. **44**, 606 (1984)
5.30 G.F. Cerofolini, G. Ferla, G.U. Pignatel, F. Riva G. Ottaviani: Thin Solid Films **101**, 263 (1983)
5.31 G.F. Cerofolini, G.U. Pignatel, E. Mazzega, G. Ottaviani: J. Appl. Phys. **58**, 2204 (1985)
5.32 D.W. Fischer, W.C. Mitchel: Appl. Phys. Lett. **45**, 167 (1984)
5.33 D.W. Fischer, W.C. Mitchel: Appl. Phys. Lett. **47**, 281 (1985)
5.34 G.F. Cerofolini, G. Ferla, G.U. Pignatel, F. Riva: Thin Solid Films **101**, 275 (1983)
5.35 M. Leung, H.D. Drew: Appl. Phys. Lett. **45**, 675 (1984)

5.36 G.F. Cerofolini, G. Ferla, G.U. Pignatel, F. Riva, F. Nava, G. Ottaviani: Thin Solid Films **109**, 137 (1983)
5.37 P. Cappelletti, G.F. Cerofolini, G.U. Pignatel: J. Appl. Phys. **54**, 853 (1983)
5.38 G.F. Cerofolini: Phil. Mag. B **47**, 393 (1983)
5.39 G.F. Cerofolini, R. Bez: J. Appl. Phys. **61**, 1455 (1987)
5.40 A. Baldereschi, J.J. Hopfield: Phys. Rev. Lett. **28**, 171 (1972)
5.41 P. Cappelletti, G.F. Cerofolini, G.U. Pignatel: Phil. Mag. A **46**, 863 (1982)
5.42 S.T. Pantelides: Appl. Phys. Lett. **50**, 997 (1987)
5.43 J.I. Pankove, P.J. Zanzucchi, C.W. Magee, G. Lucovsky: Appl. Phys. Lett. **46**, 421 (1985)
5.44 M. Stavola, S.J. Pearton, J. Lopata, W.C. Dautremont-Smith: Appl. Phys. Lett. **50**, 1086 (1987)
5.45 J.I. Pankove, C.W. Magee, R.O. Wance: Appl. Phys. Lett. **47**, 748 (1985)
5.46 N.M. Johnson: Phys. Rev. B **31**, 5525 (1985)
5.47 R.N. Hall: Phys. Rev. **87**, 387 (1952)
5.48 W. Shockley, W.T. Read: Phys. Rev. **87**, 835 (1952)
5.49 W. Shockley: Bell System Tech. J. **28**, 435 (1949)
5.50 C.T. Sah, R.N. Noyce, W. Shockley: Proc. IRE **45**, 1228 (1957)
5.51 H.J. Queisser: Solid St. Electron. **21**, 1495 (1978)
5.52 P.T. Landsberg: Proc. Roy. Soc. (London) **A 331**, 103 (1972)
5.53 J.G. Fossum, M.A. Shibib: IEEE Trans. Electron Devices **ED-28**, 1018 (1981).
5.54 G.F. Cerofolini, M.L. Polignano: J. Appl. Phys. **55**, 579 (1984)
5.55 M.J.J. Theunissen, F.J. List: Solid State Electron. **28**, 417 (1985)
5.56 E. Landi, S. Solmi: Solid State Electron. **29**, 1181 (1986)
5.57 G.F. Cerofolini: Phys. Status Solidi (a) **102**, 345 (1987)
5.58 G.F. Cerofolini, M.L. Polignano: Phys. Status Solidi (a) **100**, 177 (1987)

Chapter 6

6.1 G. Das: Mat. Res. Soc. Symp. Proc. **14**, 87 (1983)
6.2 P.M. Fahey: *Thesis*, Stanford University, Stanford, CA (1985)
6.3 M. Lannoo, J. Bourgoin, *Point Defects in Semiconductors*, Springer Ser. Solid-State Sci., **22** (Springer, Berlin, Heidelberg, 1981)
6.4 W. Zulehner, D. Huber: Crystals **8**, 1 (1982)
6.5 K. Taniguchi, D.A. Antoniadis, Y.Matsushita: Appl. Phys. Lett. **42**, 961 (1983)

6.6 R. Car, P.J. Kelly, A. Oshiyama, S.T. Pantelides: Phys. Rev. Lett. **52**, 1854 (1984)
6.7 R.B. Fair: In *Semiconductor Silicon 1977*, ed. by H.R. Huff, E. Sirtl (The Electrochem. Soc., Princeton NJ 1977) p. 968
6.8 D.A. Antoniadis, R.W. Dutton: IEEE J. Solid State Circuits **SC - 14**, 4122 (1979)
6.9 R.W. Dutton: IEEE Trans. Electron Dev. **ED-30**, 968 (1983)
6.10 W.G. Allen, K.V. Anand: Solid St. Electron. **14**, 397 (1971)
6.11 G. Masetti, S. Solmi, G. Soncini: Solid St. Electron. **16**, 1419 (1973)
6.12 G. Masetti, S. Solmi, G. Soncini: Solid St. Electron. **19**, 545 (1976)
6.13 D.A. Antoniadis, A.G. Gonzales, R.W. Dutton: J. Electrochem. Soc. **125**, 813 (1978)
6.14 K. Taniguchi, K. Kurosawa, M. Kashiwagi: J. Electrochem. Soc. **127**, 2243 (1980)
6.15 R.B. Fair: J. Electrochem. Soc. **128**, 1360 (1981)
6.16 A. Seeger, K.P. Chik: Phys. Status Solidi (A) **29**, 455 (1968)
6.17 T.Y. Tan, U. Gösele: Appl. Phys. A **37**, 1 (1985)
6.18 S.M. Hu: In *VLSI Science and Technology 1985*, ed. by W.M. Bullis, S. Broydo (The Electrochem. Soc., Pennington NJ 1985) p. 465
6.19 S.P. Murarka: J. Appl. Phys. **48**, 5020 (1978)
6.20 G.F. Cerofolini, M.L. Polignano: In *The Physics of VLSI*, ed. by J.C. Knights (Am. Inst. Phys., New York 1984) p. 225
6.21 C. Claeys, H. Bender, G. Declerck, J. Van Landuyt, R. Van Overstraeten, S. Amelinckx: In *Aggregation Phenomena of Point Defects in Silicon*, ed. by E. Sirtl, J. Gorissen (The Electrochem. Soc., Pennington NJ 1984) p. 74
6.22 G.F. Cerofolini, G. Ferla, G.U. Pignatel, F. Riva, F. Nava, G. Ottaviani: Thin Solid Films **109**, 137 (1983)
6.23 G.F. Cerofolini, R. Bez: J. Appl. Phys. **61**, 1455 (1987)
6.24 R. Baron, M.H. Young, J.K. Neeland, O.J. Marsh: Appl. Phys. Lett. **30**, 594 (1977)
6.25 M.W. Scott: Appl. Phys. Lett. **32**, 540 (1978)
6.26 R. Baron, J.P. Baukus, S.D. Allen, T.C. McGill, H. Kimura, H.V. Winston, O.J. Marsh: Appl. Phys. Lett. **34**, 257 (1979)
6.27 C.E. Jones, D. Schafer, W. Scott, R.J. Hager: J. Appl. Phys. **52**, 5148 (1981)
6.28 C.W. Searle, M.C. Ohmer, P.M. Hemenger: Solid State Commun. **44**, 1597 (1982)
6.29 C.W. Searle, P.M. Hemenger, M.C. Ohmer: Solid State Commun. **48**, 995 (1983)
6.30 G.F. Cerofolini: Phil. Mag. B **47**, 393 (1983)

Chapter 7

7.1 W.R. Wilcox, T.J. La Chapelle, D.H. Forbes: J. Electrochem Soc. **111**, 1377 (1964)
7.2 L. Baldi, G.F. Cerofolini, G. Ferla, G. Frigerio: Phys. Status Solidi (A) **48**, 523 (1978)
7.3 J.J. Burton, E.S. Machlin: Phys. Rev. Lett. **37**, 1433 (1976)
7.4 W.F. Tseng, T. Koji, J.W Mayer, T.E. Seidel: Appl. Phys. Lett. **33**, 442 (1978)
7.5 A.S. Salih, H.J. Kim, R.F. Davis, G.A. Rozgonyi: Appl. Phys. Lett. **46**, 419 (1985)
7.6 R.B. Fair: J. Appl. Phys. **50**, 860 (1979)
7.7 R.B. Fair, J.C.C. Tsai: J. Electrochem Soc. **124**, 1107 (1977)
7.8 Y. Shiraki: J. Vac. Sci. Technol. B **3**, 725 (1985)
7.9 D. Nobili: In *Properties of Silicon*, ed. by EMIS (The Institute of Electrical Engineers, to be published)
7.10 A. Armigliato, D. Nobili, P. Ostoja, M. Servidori, S. Solmi: In *Semiconductor Silicon 1977*, ed. by H.R. Huff, E. Sirtl (The Electrochem. Soc., Princeton NJ 1977), p. 638
7.11 G. Masetti, D. Nobili, S. Solmi: In *Semiconductor Silicon 1977*, ed. by H.R. Huff, E. Sirtl (The Electrochem. Soc., Princeton NJ 1977), p. 648
7.12 D. Nobili, A. Armigliato, M. Finetti, S. Solmi: J. Appl. Phys. **53**, 1484 (1982)
7.13 J.L. Hoyt, J.F. Gibbons: Mat. Res. Soc. Symp. Proc. **52**, 15 (1986)
7.14 R. Angelucci, A. Armigliato, E. Landi, D. Nobili, S. Solmi: *ESSDERC '87*, ed. by P.U. Calzolari, G. Soncini (North Holland, Amsterdam 1987) p. 461
7.15 M. Servidori, A. Armigliato: J. Mater. Sci. **10**, 306 (1975)
7.16 A. Armigliato, D. Nobili, M. Servidori, S. Solmi: J. Appl. Phys. **47**, 5489 (1977)
7.17 D. Nobili, A. Armigliato, M. Finetti, S. Solmi: J. Appl. Phys. **53**, 1484 (1982)
7.18 M. Servidori, C. del Monte, Q. Zini: Phys. Stat. Sol. (a) **80**, 277 (1983)
7.19 A. Armigliato, P. Werner: Ultramicroscopy **15**, 61 (1984)
7.20 A. Armigliato, A. Bourret, S. Frabboni, A. Parisini: Inst. Phys. Conf. Ser. **87**, 55 (1987)
7.21 S. Fischler: J. Appl. Phys. **33**, 1615 (1962)
7.22 H. Statz: J. Phys. Chem. Solids **24**, 699 (1963)
7.23 F.A. Trumbore: Bell System Tech. J. **39**, 205 (1960)
7.24 P. Cappelletti, G.F. Cerofolini, G.U. Pignatel: Phil. Mag. A **46**, 863 (1982)

7.25 P. Cappelletti, G.F. Cerofolini, G.U. Pignatel: Phil. Mag. A **47**, 623 (1983)
7.26 A.A. Maradudin, E.W. Montroll, G.H. Weiss: *The Theory of Lattice Dynamics in the Harmonic Approximation* (Academic, New York 1963)
7.27 A. Baldereschi, J.J. Hopfield: Phys. Rev. Lett. **28**, 171 (1972)

Chapter 8

8.1 F.J. Grunthaner, P.J. Grunthaner: Materials Sci. Repts. **1**, 65 (1986)
8.2 J.S. Johannesen, W.E. Spicer, Y.E. Strausser: J. Appl. Phys. **47**, 3028 (1976)
8.3 C.R. Helms, Y.E. Strausser, W.E. Spicer: Appl. Phys. Lett. **33**, 767 (1978)
8.4 C.R. Helms, N.M. Johnson, S.A. Schwarz, W.E. Spicer: J. Appl. Phys. **50**, 1067 (1979)
8.5 B.E. Deal: IEEE Trans. Electron Dev. **ED-27**, 606 (1980)
8.6 A. Goetzberger, E. Klausmann, M.J. Schulz: CRC Critical Reviews in Solid State Sciences **1**, 1 (1976)
8.7 B.E. Deal: J. Electrochem. Soc. **121**, 198C (1974)
8.8 C.N. Berglund: IEEE Electron Dev. **ED-13**, 701 (1966)
8.9 G.F. Cerofolini, G. Ferla, G. Spadini: Thin Solid Films **68**, 315 (1980)
8.10 S.I. Raider, A. Berman: J. Electrochem. Soc. **125**, 629 (1978)
8.11 P.J. Caplan, E.H. Poindexter, B.E. Deal, R.R. Razouk: J. Appl. Phys. **50**, 5847 (1979)
8.12 E.H. Poindexter, P.J. Caplan, B.E. Deal, R.R. Razouk: J. Appl. Phys. **52**, 879 (1981)
8.13 Y. Nishi: Jap. J. Appl. Phys. **5**, 333 (1966)
8.14 Y. Nishi: Jap. J. Appl. Phys. **10**, 52 (1971)
8.15 B.E. Deal, A.S. Grove: J. Appl. Phys. **36**, 3770 (1965)
8.16 G. Mende, J. Finster, D. Flamm, D. Schulze: Surface Sci. **128**, 169 (1983)
8.17 S.M. Hu: Appl. Phys. Lett. **42**, 872 (1983)
8.18 S.M. Hu: J. Appl. Phys. **55**, 4095 (1984)
8.19 G.F. Cerofolini: In *Colloid Science*, ed. by D.H. Everett (The Chemical Soc., London 1983) vol. IV, p. 59.
8.20 G.F. Cerofolini: Z. Phys. Chem. (Leipzig) **259**, 1020 (1978)
8.21 H.Z. Massoud, J.D. Plummer, E.A. Irene: J. Electrochem. Soc. **132**, 2685 (1985)
8.22 H.Z. Massoud, J.D. Plummer, E.A. Irene: J. Electrochem. Soc. **132**, 2693 (1985)

8.23 P.M. Fahey, R.W. Dutton, M. Moslehi: Appl. Phys. Lett. **43**, 683 (1983)
8.24 P.M. Fahey, G. Barbuscia, M. Moslehi, R.W. Dutton: Appl. Phys. Lett. **46**, 784 (1985)
8.25 P.M. Fahey: *Thesis* (Stanford University, Stanford, CA, 1985)
8.26 S.M. Hu: J. Appl. Phys. **57**, 1069 (1985)

Chapter 9

9.1 H. Strack, K.R. Mayer, B.O. Kolbesen: Solid State Electron. **22**, 135 (1979)
9.2 J.M. Dishman, S.E. Haszko, R.B. Marcus, S.P. Murarka, T.T. Sheng: J. Appl. Phys. **50**, 2689 (1979)
9.3 S.P. Murarka, T.E. Seidel, J.V. Dalton, J.M. Dishman, M.H. Read: 156th Meeting Electrochem. Soc. (1979), Abs. No. 489
9.4 H.R. Huff, F. Shimura: SEMI Symp. Low Temperature Processing for VLSI Large Diameter Wafers (1984)
9.5 Y. Kondo: In *Semiconductor Silicon 1977*, ed. by H.R. Huff, R.J. Kriegler, Y. Takeishi: (The Electrochem. Soc., Pennington NJ 1981) p. 220
9.6 S. Prussin, S.P. Li, R.H. Cockrum: J. Appl. Phys. **48**, 4613 (1977)
9.7 G.A. Rozgonyi, P.M. Petroff, M.H. Read: J. Electrochem. Soc.**122**, 1725 (1975)
9.8 P.M. Petroff, G.A. Rozgonyi, T.T. Sheng: J. Electrochem. Soc. **122**, 565 (1975)
9.9 S.P. Murarka: J. Appl. Phys. **48**, 5020 (1978)
9.10 F. Shimura, R.A. Craven: In *The Physics of VLSI*, ed. by J.C. Knights (Am. Inst. Phys., New York 1984) p. 205
9.11 T.Y. Tan, E.E. Gardner, W.K. Tice: Appl. Phys. Lett. **30**, 175 (1977)
9.12 K. Nagasawa, Y. Matsushita, S. Kishino: Appl. Phys. Lett. **37**, 622 (1980)
9.13 K. Yamamoto, S. Kishino, Y. Matsushita, T. Izuka: Appl. Phys. Lett. **36**, 195 (1980)
9.14 H.H. Busta, H.A. Waggener: J. Appl. Phys. **48**, 4385 (1977)
9.15 H.H. Busta, H.A. Waggener: J. Electrochem. Soc. **124**, 1424 (1977)
9.16 R.L. Meek, T.E. Seidel: J. Phys. Chem. Solids **36**, 731 (1975)
9.17 A. Goetzberger, W. Shockley: J. Appl. Phys. **31**, 1821 (1960)
9.18 L. Baldi, G.F. Cerofolini, G. Ferla, G. Frigerio: Phys. Status Solidi (a) **48**, 523 (1978)
9.19 L. Baldi, G.F. Cerofolini, G. Ferla: 154th Meeting Electrochem. Soc. (1978) abs. 209

9.20 L. Baldi, G.F. Cerofolini, G. Ferla: J. Electrochem. Soc. **127**, 125 (1980)
9.21 G.F. Cerofolini, G. Ferla: In *Semiconductor Silicon 1981*, ed. by H.R. Huff, R.J. Kriegler, Y. Takeishi (The Electrochem. Soc., Pennington NJ 1981) p. 724
9.22 G.F. Cerofolini, M.L. Polignano: J. Appl. Phys. **55**,579 (1984)
9.23 P. Cappelletti, G.F. Cerofolini, M.L. Polignano: J. Appl. Phys. **57**, 1406 (1985)
9.24 G.F. Cerofolini, G. Ferla: 156th Meeting Electrochem. Soc. (1979) abs. 492
9.25 L. Baldi, G.F. Cerofolini, G. Ferla: Surf. Technol. **8**, 161 (1979)
9.26 W.F. Tseng, T. Koji, J.W Mayer, T.E. Seidel: Appl. Phys. Lett. **33**, 442 (1978)
9.27 G.F. Cerofolini, M.L. Polignano, H. Bender, C. Claeys: Phys. Status Solidi (a) **103**, 643 (1987)
9.28 G.B. Bronner, J. Plummer: 166th Meeting Electrochem. Soc. (1984) abs. 483
9.29 R.J. Falster, D.N. Modlin, W.A. Tiller, J.F. Gibbons: J. Appl. Phys. **57**, 554 (1985)
9.30 R. Falster: Appl. Phys. Lett. **46**,737 (1985)
9.31 G.B. Bronner, J.D. Plummer: J. Appl. Phys. **61**, 5286 (1987)
9.32 U. Gösele, W. Frank, A. Seeger: Appl. Phys. A **23**, 361 (1980)

Chapter 10

10.1 W. Shockley: IEEE Trans. Electron Dev. **ED-23**, 597 (1976)
10.2 L. Derick, C.J. Frosch: U.S. Patent No. 2,802,760 (1957)
10.3 J. Andrus: U.S. Patent No. 3,122,817 (1964)
10.4 J.A. Hoerni: U.S. Patent No. 3,025,589 (1962)
10.5 R.N. Noyce: U.S. Patent No. 2,981,877 (1961)
10.6 J.E. Lilienfeld: U.S. Patent No.1,745,175 (1930)
10.7 O. Heil: British Patent No. 439,457 (1935)
10.8 D. Khang: M.M. Atalla: IRE Solid State Device Research Conf. (1960)
10.9 J.C. Sarace, R. Kerwin, D.L. Klein, R. Edwards: Solid State Electron. **11**, 653 (1968)
10.10 F. Faggin, T. Klein, L. Vadasz: IEEE Intl. Electron Device Meeting (1968)
10.11 L.L. Vadasz, A.S. Grove, T.A. Rowe, G.E. Moore: IEEE Spectrum **6**, 28 (1969)
10.12 J.A. Appels, E. Kooi, M.M. Paffen, J.J. Schatorjé, W.H.C.G. Verkuijlen: Philips Res. Repts. **25**, 118 (1970)

10.13 J.A. Appels, M.M. Paffen: Philips Res. Repts. **26**, 157 (1971)
10.14 E. Kooi, J.G. van Lierop, W.H.C.G. Verkijlen, R. de Werdt: Philips Res. Repts. **26**, 166 (1971)
10.15 F. Morandi: IEEE Intl. Electron Device Meeting (1969)
10.16 W. Shockley: U.S. Patent No. 2,787,564 (1957)
10.17 B.E. Deal, J.M. Early: J. Electrochem. Soc. **126**, 20C (1979)
10.18 H.K.J. Ihantola, J.L. Moll: IEEE Trans. Electron Dev. **ED-11**, 324 (1964)
10.19 R.H. Dennard, F.H. Gaensslen, H.N. Yu, V.L. Rideout, E. Bassous, A. LeBlanc: IEEE J. Solid St. Circuits **SC-9**, 256 (1974)
10.20 G. Baccarani, M.R. Wordeman, R.H. Dennard: IEEE Trans. Electron Dev. **ED-31**, 452 (1984)
10.21 T.H. Ning: In *The Physics of VLSI*, ed. by J.C. Knights (Am. Inst. Phys., New York 1984) p. 151
10.22 J.C. Bean: In *The Physics of VLSI*, ed. by J.C. Knights (Am. Inst. Phys., New York 1984) p. 198
10.23 R.J. Chater, J.A. Kilner, E. Scheid, S. Cristoloveneau, P.L.F. Hemment, K.J. Reeson: Appl. Surf. Sci. **30**, 390 (1987)
10.24 S.M.S. Liu, C.H. Fu, G.F. Atwood, H. Dun, J. Langston, E. Hazani, E.Y. So, S. Sachdev, K. Fuchs: IEEE J. Solid St. Circuits **SC-17**, 810 (1982)
10.25 M.L. Polignano, G.F. Cerofolini, H. Bender, C. Claeys, J. Reffle: In *ESSDERC 87*, ed. by P.U. Calzolari, G. Soncini (North Holland, Amsterdam 1987) p. 335
10.26 J. Hui, T.Y. Chiu, S. Wong, W.G. Oldham: IEEE Electron Dev. Lett. **EDL-2**, 244 (1981)
10.27 K.Y. Chiu, J.L. Moll, J. Manoliu: IEEE Trans. Electron Dev. **ED-29**, 536 (1982)
10.28 J.C. Hui, T.Y. Chiu, S.S. Wong, W.G. Oldham: IEEE Trans. Electron Dev. **ED-29**, 554 (1982)
10.29 N. Hashimoto, Y. Yatsuda, S. Mutoh: Jap. J. Appl. Phys. **16**, Suppl. 1, 73 (1977)
10.30 M.L. Polignano, G.F. Cerofolini, H. Bender, C. Claeys: In *ESPRIT '85*, ed. by the Commission of the European Communities (North Holland, Amsterdam 1986) Part 1, p. 233
10.31 L.A. Berthoud: Microelectron. and Reliab. **16**, 165 (1977)
10.32 S.P. Murarka: J. Vacuum Sci. Technol. **17**, 775 (1980)
10.33 G. Ottaviani, J.W. Mayer: In *Reliability and Degradation*, ed. by M.J. Howes and D.V. Morgan (Wiley, New York 1981) p. 105
10.34 T.T. Sheng, R.B. Marcus: J. Electrochem. Soc. **128**, 881 (1981)
10.35 R.W. Keyes: Science **195**, 1230 (1977)
10.36 O.G. Folberth: IEEE J. Solid St. Circuits **SC-16**, 51 (1981)
10.37 C. Svensson: Integration VLSI J. **1**, 3 (1983)
10.38 P.M. Solomon: In *The Physics of VLSI*, ed. by J.C. Knights (Am. Inst. Phys., New York 1984) p. 172.

10.39 G.E. Moore: *Technical Digest 1975 International Electron Devices Meeting* (1975) p. 11
10.40 A.D. Wilson: *The Physics of VLSI*, ed. by J.C. Knights (Am. Inst. Phys., New York 1984) p. 69
10.41 J.C. Bean, E.A. Sadowski: J. Vacuum Sci. Technol. **20**, 137 (1982)
10.42 G.H. Dohler: Physica Scripta **24**, 430 (1981)
10.43 S.S. Iyer, R. A. Metzger, F.G. Allen: J. Appl. Phys. **52**, 5608 (1981)
10.44 H. Ishiwara, T. Asano: Appl. Phys. Lett. **40**, 66 (1982)
10.45 H. Zogg, S. Blunier: Appl. Surf. Sci. **30**, 402 (1987)
10.46 A.H. Van Ommen, M.P.A. Viegers: Appl. Surf. Sci. **30**, 383 (1987)
10.47 Several authors: 172th Meeting Electrochem. Soc. (1987) abs. no. 326-394
10.48 G.F. Knoll: Nucl. Instrum. Meth. B **24/25**, 1021 (1987)
10.49 S.H. Moseley, J.C. Mather, D. McCammon: J. Appl. Phys. **56**, 1257 (1984)
10.50 D. McCammon, S.H. Moseley, J.C. Mather, R.F. Mushotzky: J. Appl. Phys. **56**, 1263 (1984)
10.51 B. Cabrera, L.M. Krauss, F. Wilczek: Phys. Rev. Lett. **55**, 25 (1985)

Acronyms and Abbreviations

a	amorphous
CCC	central cell correction
CZ	Czochralski
DAT	donor-acceptor twin
d.c.	diamond cubic
DDD	deep dopant description
d.h.	diamond hexagonal
DZ	denuded zone
e	electron
EG	external gettering
EMA	effective mass approximation
EMIS	Electronic Materials Information Service
ESF	extrinsic stacking fault
ESR	electron spin resonance
FZ	float–zone
h	hole
HI	high temperature (IG process)
HMOS-III	a third-generation MOS technology
HREM	high resolution electron microscopy
HSI	horrendus scale integration
i	interstitial atom
I^2	ion implantation
IEEE	Institute of Electrical and Electronic Engineers
IG	internal gettering
INTEL	an integrated circuit company in the Silicon Valley
ISF	intrinsic stacking fault
LAMEL	laboratory of electronic materials (an Italian National Research Council laboratory)
LO	low temperature (IG process)
LOCOS	local oxidation of silicon
LSI	large scale integration
MOS	metal-oxide-semiconductor
MSI	medium scale integration
OSF	oxidation stacking fault

PLANOX	planar oxidation
RBS	Rutherford backscattering spectroscopy
SEM	scanning electron microscopy
SF	stacking fault
SGS	Societá Generale Semiconduttori (an Italian Company)
SPE	solid phase epitaxy
SRH	Shockley–Read–Hall
SSI	small scale integration
STSD	standard theory of shallow dopants
TEM	transmission electron microscopy
ULSI	ultra large scale integration
v	vacancy
VLSI	very large scale integration

Subject Index